THE ATOMS WITHIN US

THE ATOMS
WITHIN US

by ERNEST BOREK

 NEW YORK 1961

COLUMBIA UNIVERSITY PRESS

For F. F. S.

ACKNOWLEDGMENTS

BIOLOGICAL CHEMISTRY is developing so rapidly that even the textbooks do not achieve a synthesis of the field but merely compile the emerging facts. I therefore appealed to several experts for help to weed out errors from this attempt at a synthesis. I list these friends and colleagues not to hide behind them if errors should be pointed out, but to thank them: Doctors Aaron Friedman, Abraham Mazur, Stanley Miller, and Stephen Zamenhof.

I am particularly indebted to my associate Miss Ann Ryan for library and editorial assistance. My colleague Miss Harriet E. Phillips designed the dust jacket.

The photographs in Chapter 14 are reprinted with the kind permission of Dr. M. F. Perutz, F.R.S., and Dr. Sydney Brenner, both of Cambridge University.

PREFACE

EVERY MAN, woman and child inhabiting this planet became an unwitting guinea pig in a vast biochemical experiment on July 16, 1945, when the first atom bomb was successfully detonated. This experiment, whose outcome—the effect of an increment of radiation on life—may not be known for centuries, serves to emphasize the profound involvement of every one of us in biochemistry. Actually no such dramatic reminder should be needed. Every living creature is a chemical conglomerate existing in a vast sea of chemicals.

Biological chemistry is the study of the molecular structure and the molecular mechanism of living things. It is at present the most rapidly growing and undoubtedly the most spectacular of the biological sciences. I have written this book to acquaint the interested reader with biochemistry—my field.

I use no chemical formulas in this book. After all, they are merely shorthand symbols for facts and ideas. The formula H-O-H merely states that a water molecule is made up of two atoms of the element hydrogen and one of the element oxygen and tells something of the energy binding the atoms together. It no more depicts the exact appearance of a water molecule than a blueprint depicts the exact appearance of the engine which will be built from it.

While describing the work of the biochemist without using

his baffling symbols, I tried to follow the precept for the exposition of a science to a lay audience given by the great physicist James Clerk Maxwell: "For the sake of persons of different types, scientific truth should be presented in different forms and should be regarded as equally scientific, whether it appears in the robust form and vivid coloring of a physical illustration, or in the tenuity and paleness of a symbolical expression."

I should emphasize that while I have made every effort to achieve "scientific truth," I have not attempted to present the development of the various areas of the biochemist's endeavor in complete historical sequence. I feel that an approach which must undoubtedly yield a very long and tedious book could not be risked in a work which is aimed not at the captive audience of students but rather at the free public. In most of the chapters that follow the historical background is sketched in only to serve as a frame within which a pattern of current concepts could be woven. In turn, only such current biochemical work as could be fitted into that pattern is included. What I have tried to achieve is a connected story, which shows the biochemist at his bench, traces the growth of ideas that guide his hands, and unfolds his view on the mechanism of the living machine.

I am not an admirer of the genre of science reporting in which the work of the scientist is but a thin filling sandwiched between soggy chunks of elaborate accounts of the real and imaginary drama of his life. ("A haggard man in white is pacing in the dim laboratory. His eyes, granules of shining coal in dark pits; his mouth a thin, tightly drawn ribbon as he mutters: 'This time it must work!' ") This is corn which has turned rancid with overuse. But even if I were an admirer of this sort of thing I could not perpetrate it in this book. Biological chemistry is a young science; many of the scientists whose work is included in the book are very much alive. Indeed, I have concentrated on the science rather than on the scientist to the point where I have not identified by name many of the contemporary biochemists—often only their work is mentioned. I have thus managed to evade the responsibility of choosing whom to include from among fellow scientists, many of whom are acquaintances and some personal friends.

I have kept predictions and fantasies about the future achievements of biochemistry to a minimum. Even with the most limited of imaginations it is easy to engage in such pastimes. It is an indoor sport during the coffee respite for many working scientists. But such speculations should be restricted to the experts in the field. For only they know the rules of the game, only they can clearly discern the difference between science fiction and a plausible goal within experimental reach. It is not necessary to impress the world with miracles of the future; we have achieved enough miracles already to prove our worth.

Some understanding of the work of the scientist by every member of a free society is an increasingly urgent need. The burden of the support of science has been gradually shifting. The alchemist was supported by the largesse of kings and princes; with the rise of capitalism and the concentration of wealth some oil and beer barons assumed the roles of benefactors; and, now, since the federal government underwrites more than half of the scientific research in our country, the taxpaying public has become our patron.

In 1669 the alchemist Hennig Brandt isolated what he thought was some magic stuff, phosphorus—it glowed in the dark!—from the phosphates of urine. He was supported—inadequately he thought—by his patron, a German prince. About two hundred and ninety years later a far more gifted alchemist, Nobel prize winner Dr. Arthur Kornberg, put together from more complex phosphates a magic stuff, DNA, which is the genetic material of living things. *His* patron was the United States Public Health Service, which is one of the dispensing agencies of the American public's patronage.

The role of the patron of the sciences is no less hazardous than that of the arts. Shoddy but spectacular undertakings can outshout research endeavors of depth and integrity. Only the well informed can be discerning. I hope this book will serve as a primer in my field for my patrons.

ERNEST BOREK

New York
November, 1960

CONTENTS

THE ATOMS WITHIN US

Science is the poetry of the intellect. . . .

LAWRENCE DURRELL

I am a little world made cunningly
Of elements, and an angelic sprite.
JOHN DONNE

1 . THE STUFF OF LIFE

A FERTILIZED EGG is at once the most precious and the most
baffling thing in the universe. Within the thin shell of a hen's
egg is locked not just a mere mass of egg white and yolk, but a
promise of a beautiful creature of flesh, blood, and bones—a
promise of the continuity of life.

The mode of fulfillment of that promise is the most baffling
mystery within the horizon of the human mind. The mystery
of the sun's energy no longer eludes us: we have converted mass
into energy; we perpetuate a Lilliputian aping of the sun with
each detonating hydrogen bomb. But the mastery of the tools,
of the energy, and of the blueprint which shapes an apparently
inert mass into life, is still but a thin hope.

A century and a quarter ago there was not even such a hope.
In our striving for knowledge we had to contend not only with
the secret ways of the universe, but often we had to surmount
man-made barriers before we could approach our quest. One
such road block in the path to knowledge was the principle of
vitalism which dominated scientific thought until the middle
of the nineteenth century. Scientists had explored with fruitful
zest the nonliving world, but they halted with awe and im-
potence before a living thing. It was believed by the vitalists
that the cell membrane shrouded mysterious vital forces and
"sensitive spirits." It was an unassailable tenet that not only

could we not fathom these mysteries, but that we should never be able to duplicate by any method a single product of such vital forces. An uncrossable chasm was supposed to separate the realm of the living, organic world and the realm of the non-living, inorganic world.

In 1828, a young man of twenty-eight unwittingly bridged that chasm. He made something in the chemical laboratory which, until then, had been made only in the body of a living thing. This achievement was the "atom bomb" of the nineteenth century. Its influence in shaping our lives is far greater than the influence of atomic energy will be.

Had there been science writers on the newspapers of that day they could very well have unfurled all the clichés of their present-day counterparts about the "significant breakthrough." But, oddly, not only were the man and his feat unknown to his contemporaries, he is practically unknown even today. The generals of the Napoleonic era—Blucher, Ney, Bernadotte—are known to many, but their contemporary, Friedrich Wöhler, who was a giant of intellect and influence compared to them, is known only to chemists.

Friedrich Wöhler was a student of medicine at Heidelberg in the early 1820s. His chemistry teacher, Gmelin, was one of those rare teachers who not only dispensed knowledge but "shaped souls." Under his guidance Wöhler left medicine and became a chemist. He more than justified his teacher's faith, for, in addition to the great discovery which shook the foundation of vitalism, we owe to Wöhler the discovery of two elements, aluminum and beryllium. After absorbing all that Heidelberg could offer, he went to Sweden to work with Berzelius, the greatest contemporary master of chemistry.

There Wöhler discovered, quite by accident, a method of making urea, a substance theretofore produced only by the cells

of living creatures. So contrary to current ideas was this achievement that he published it only after numerous repetitions, four
years later. How did he learn the secret of the sensitive spirits?
How did he make urea?

Urea is a substance found as a waste product in the urine
of some animals. It can be cajoled out in the form of pure
white crystals, by the knowing hands of the chemist. Like all
other pure compounds, urea has characteristic attributes which
distinguish it from any other substance. Sugar and salt are two
different pure compounds which superficially look alike. But
the tongue tells them apart with unfailing ease. Their different
impact on our discerning taste buds is but one of many differences between salt and sugar. The chemist has discovered scores
of other differences. The elements of which they are composed,
the temperature at which they melt, certain optical properties
of the crystals, these are some of the distinguishing lines in the
fingerprint of a compound, by means of which the chemist can
recognize a particular one among the multitudes.

Urea happens to be made of four different elements. One
atom of carbon, one of oxygen, two of nitrogen, and four of
hydrogen make up a urea molecule. These atoms are attached
to each other in a definite pattern, a pattern unique to a single
substance, urea. The force which binds these atoms into the
pattern of urea is the energy in the electrons of those atoms.
A chemical union between atoms is a light, superficial affair.
Two atoms meet, some of their outer electrons become en-
tangled, and a temporary union is formed. The nucleus of the
atom is completely unaffected by a chemical union. The energy
involved in the making or breaking of such a union is minuscule
compared to the vast energy locked within the nucleus of the
atom—the monstrous nuclear, or atomic, energy.

Until Wöhler succeeded in making it, urea could be fashioned only in a living animal by unknown, awesome, animated spirits. The spirits whimsically flung out their product into the urine for reasons, it was thought, no human could fathom.

Wöhler had no intention of making urea in a test tube. He was studying a simple, undramatic, chemical reaction. He wanted to make a new inorganic compound, ammonium cyanate. He went through a variety of manipulations which were expected to yield the new substance. As the final step, he boiled away the water and some white crystals were left behind. But they were not the new inorganic salt that he had expected; they were the very same urea which animals excrete.

It so happens that in ammonium cyanate, the substance Wöhler had set out to make, there are the same elements in the same number as in urea. The difference is the pattern the atoms form. The pattern of ammonium cyanate was disrupted by the heat of the boiling water and the atoms rearranged themselves to form urea. (The changing of chemical structures by heat is not unusual; indeed it is an everyday household feat —a boiled egg is quite different from an uncooked one.)

Simple? It looks simple now, 130 years later, when animated spirits have become scientific antiques and the manufacture of synthetic vitamins is a big industry. Thus, only in our hindsight, which sharpens with the years elapsed, does it look simple.

The importance of the finding was not lost to Wöhler and some of his contemporaries. While he wrote very modestly of his achievement in his four-page technical communication, he let himself go when writing to his mentor, Berzelius. "I must now tell you," he wrote, "that I can make urea without calling on my kidneys, and indeed, without the aid of any animal, be it man or dog."

And the master replied graciously. "Like precious gems—alu-

minum and artificial urea—two very different things coming
so close together [1]—have been woven into your laurel wreath."

The homage of history was paid by Sir Frederick Gowland
Hopkins, Nobel Prize winner, on the centenary of the discovery: "The intrinsic historic importance of Wöhler's synthesis can hardly be exaggerated. So long as the belief held
ground that substances formed in the plant or animal could
never be made in the laboratory, there could be no encouragement for those who instinctively hoped that chemistry might
join hands with biology. The very outermost defences of vitalism seemed unassailable."

Though the barrier between organic and inorganic chemistry was demolished by Wöhler, the terms have been retained,
but with new meanings. Organic chemistry now embraces the
chemistry of the compounds of carbon. This element is uniquely
fecund in forming compounds. Over a million different compounds of carbon have been made by organic chemists. And
new ones, by the dozen, are being added daily.

If one feels, as many of us do sometimes, that we are being
drowned in those cataracts of compounds, solace can be had
in the knowledge that even Wöhler felt inundated. Over a
hundred years ago he wrote: "Organic Chemistry nowadays almost drives one mad. To me it appears like a primeval tropical
forest full of the most remarkable things; a dreadful endless
jungle into which one does not dare enter, for there seems to
be no way out."

In that "jungle" of compounds are hidden veritable mines
of drugs, perfumes and plastics. It is impossible to predict
what uses a new compound may have. Sulfanilamide, DDT,

[1] Wöhler discovered the new element aluminum in 1827 and announced
the synthesis of urea in 1828.

four members of the vitamin B group, and many other drugs lay for years on the shelf of organic chemistry before their potency in nutrition and medicine was discovered.

To inorganic chemistry is relegated the study of the compounds of the other elements. All of these together number only a paltry 50,000.

Encouraged by Wöhler's tremendous success, other chemists boldly set out in search of the products and components of the living cell. Their tools and training proved so apt for the task that we have learned more about the structure of life in the thirteen decades since Wöhler's achievement than in all of recorded history. Hundreds of compounds which had been pried out from the cell were duplicated [2] in the laboratory after Wöhler's fashion.

These successes changed medicine, changed nutrition, changed our way of life. Are you eating vitamin-enriched bread? Were you given large doses of vitamin K before an operation to stop excessive hemorrhage? Are you receiving injections of hormones? Has the life of a dear one been saved by penicillin? For all these bounties, thank Wöhler and the generations of chemists his deed encouraged.

Paralleling the chemist's search for the molecular components of the cell biologists were charting the landmarks visible under a microscope in that Lilliputian universe. The universe is so small that, for example, it would take about five thousand human red blood cells to cover the dot over the letter i on this page. The boundary of this speck of jelly is the cell membrane,

---

[2] The term *synthetic* has acquired an opprobrious meaning, indicating a poor substitute for the genuine. However, when the chemist synthesizes a known substance he makes an identical duplicate of what nature had made. Vitamin C extracted from an orange or from the vat of the chemical manufacturer is exactly the same material; it is impossible to tell the products apart by any means.

which in plants and bacteria is buttressed by a more substantial cell wall. The cell membrane is not merely an inert bag to contain the minuscule watery world. It may well be the most active unit of the cell's many components. The membrane sorts out with uncanny certainty the hundreds of different substances swimming around it and permits or denies passage according to the cell's needs. Inside the cell the microscope reveals a startling variety of structures. In most cells there is a central globule, the nucleus—an inner fortress as it were, guarded by its own membrane. This tiny sphere is crisscrossed by flimsy, lacy threads and more robust strands. These strands, called the chromatins because they absorb certain dyes, perform a veritable ballet of convolutions, separations, and realignments prior to the division of a cell into new ones. The symbolism of this ballet is more arcane than that of the most *avant garde* of modern dances. The correlation of the restless patterns of the chromatin granules with cell division merely from viewing static images under a microscope is a monument to the imagination of the classical biologist.

The landscape between the nucleus and the cell's periphery, the so-called cytoplasm, abounds in a variety of structures. Small, compact hillocks—the mitochondria—and still smaller mounds—the microsomes—are embedded apparently at random. Here and there lacy filaments—the Golgi bodies—wind a lazy path. In some cells still other features—pigments, starch granules, globules of fat—dot the landscape. Variety is the rule, uniformity the exception in the world of the cells.

Classical biologists have deduced, with piercing ingenuity, many of the functions of the components of the cell. They recognized that the biological individuality of a living organism must be bestowed on its offspring through the structural units of the cells.

But how do the various parts of a cell perform their multitude of roles? If the cell is the unit of structure of life, does it have a unit of function? The microscope could give us no answer to this question, for it can reveal to us only the rigid profile of a static world. From such an image we had no idea of the maelstrom of activity going on with explosive speed and boundless variety beyond the prying reach of the best microscope. New tools, new ideas were needed.

These were provided by the chemists who created a new discipline and assembled a new body of knowledge: Biological Chemistry.

In this chronicle of the achievements of those chemists we may be carried away by our enthusiasm and pride in our ever-growing prowess and knowledge. We should at such times recall what the discoverer of the circulation of blood said in 1625. "All we know," said William Harvey, "is still infinitely less than all that still remains unknown." That humble statement is as true today as it was then, for 335 years is a very short time to study anything as wonderfully complex as a living cell.

What have the generations of biochemists found in living things? First of all, they struck water. Lots of water. About 70 percent of the human body is water. "Water thou art and to water returnest" would be a chemically more accurate, if less euphonious, description of our corporal denouement. The amount of water in the human body is surprisingly constant. When it increases locally in a small area the tissues become swollen. There is a general increase in the water content of tissues in old age. The shrunken, externally parched appearance of old age is misleading, for the water content of the body is actually increased. Whether this increased hydration has any causal relation to aging is one of the multitude of unanswered questions which makes Harvey's humble statement all too true.

We must not find in the hydration of old age justification for replacing water as a staple beverage by more potent fluids. Alcohol actually introduces more water into the body than a similar weight of water does. An ounce of absolute, 200-proof alcohol will produce about one and one sixth ounces of water. The formation of that much water, although surprising, is perfectly possible. One of the constituent elements of alcohol is hydrogen. When alcohol is burned in the body the hydrogen is combined with oxygen that we inhale from the atmosphere, to form water.

This type of water formation is not unique to alcohol; every food is a source of water in a living creature. The camel puts this bit of biochemistry to a very practical use. To cross the desert he needs both food and water. The camel's hump, which is largely fat, provides both. Assuming that a camel's hump contains one hundred pounds of fat, the camel will derive from burning that fat huge amounts of energy—over 400,000 Calories—and, as a bonus, fifty quarts of water.

Water serves us well: it is a freight canal for the transport of foods and wastes; it regulates the body temperature by evaporating as cooling is needed; it is the remarkably efficient lubricating fluid for the body's many joints; and, finally, it makes up 70 percent of the human body. We living creatures contain only 30 percent solids. We have about the same proportion of solids as a cup of water containing ten teaspoonfuls of sugar. Why don't we flow as freely as that sugar solution does? Why is our flesh, in Hamlet's words, "too, too solid"? Why doesn't it "melt, thaw and resolve itself into a dew"?

The biochemist's prosaic answer to the prince would be: "Because we have proteins."

Proteins are able to bind large amounts of water into themselves, forming semisolid jellies. Anyone who has ever made gelatin dessert knows this. A small amount of a dry protein—gelatin—soaks up a large volume of water and produces a semirigid mass. Not all of our proteins bind water as readily as gelatin does; we must have considerable amounts of free, unbound water in our bodies. But such water is usually confined within tubes or tissues and thus we can retain the body's characteristic solidity.

Proteins are the most characteristic components of the cell; all of life's processes are tied up in them. The term protein has been very aptly devised. It means "of primary importance."

We find proteins in every cell. Egg white, cheese, hair, and nails are composed largely of proteins. If a protein is boiled with strong acid for twenty hours it loses its identity and its characteristic properties. The edifice of the protein molecule is crumbled into its component bricks by the hot acid. The bricks that we can find in a solution of a dismantled protein are the amino acids.

There are more than twenty different amino acids. They all contain the element nitrogen. Hence the great need for nitrogenous fertilizers to insure good crops. Plants need the nitrogen from the soil to fashion their amino acids and proteins.

As nitrogen is the characteristic element in all proteins, phosphorus serves the same function in another large class of cellular components, the nucleic acids. (They were once thought to be restricted to the nucleus of cells—hence their name—but actually we find them throughout the cell.)

What else, besides water, proteins, and nucleic acids does the biochemist find in the cell? He finds fats and sugars, the two other large components which make up the solid matter of the cell. (All of these will be subjects of discussion in later chapters.)

In addition, the cell contains, in minute amounts, scores of other organic substances, such as vitamins and hormones. It also contains a large number of inorganic salts. Some of the salts are present in considerable amounts; of others we find but traces.

Water, proteins, nucleic acids, fats, sugars, and salts, such are the mundane substances the chemist finds within a living cell. Certainly these are not "such stuff as dreams are made on." And yet, on such stuff is built the edifice of the improbable dream that is life. All the greater, therefore, is the miracle of that life.

O! it is excellent
To have a giant's strength,
SHAKESPEARE

2 . ENZYMES

GIANT MOLECULES
WITH GIGANTIC KNOW-HOW

WE LIVE because we have enzymes. Everything we do—walking, thinking, reading these lines—is done with some enzyme process. Life may be defined as a system of integrated, cooperating enzyme reactions.

The best way to get acquainted with an enzyme is to observe it in action. In order to break down a protein, say some egg white, into its constituent amino acids in the laboratory, we must use rather drastic methods. We add ten times its weight of concentrated acid and boil the mixture for twenty hours. In the stomach and small intestine the same disintegration of the egg takes place in a couple of hours at body temperature and without such strong acid. This chemical sleight of hand is performed by the enzymes made for this purpose by the cells in the stomach wall. If we add the stomach lining of a recently slaughtered hog to boiled egg white and keep the mixture at body temperature for a few hours, the egg white will be disintegrated into its amino acids just as effectively as it would have been in the stomach of the live animal.

Their enormous activity is characteristic of enzymes. One

ounce of an enzyme preparation from the hog's stomach will digest 50,000 ounces of boiled egg white in two hours. The same preparation will also clot milk—it is the active ingredient of rennet powders. The potency of enzymes can be demonstrated even more impressively: one ounce will clot 2,800,000 quarts of milk.

It is astonishing how recent is our knowledge of biological mechanisms. Until the early 1820s we had no conception of what happens to the food in our mysterious interiors. Prior to that there were a host of conjectures, not the least imaginative of which was that a band of little demons was busily engaged in our stomachs, macerating our food.

Toward the end of the eighteenth century Réaumur in France and Spallanzani in Italy took the first steps toward the exploration of the stomachs of animals. They fed food, enclosed in a wire cage or in a perforated capsule, to animals and, at various intervals, retrieved the containers by means of attached strings. They noted the dissolution of the food but could not even guess at the nature of the substances which were responsible for these changes.

A whole set of new explanations was brought forth, but that they did not gain universal acceptance is obvious from the irascible fragment of a lecture by the English physician William Hunter: "Some physiologists will have it, that the stomach is a mill, others, that it is a fermenting vat, others, again that it is a stew pan; but in my view of the matter, it is neither a mill, a fermenting vat, nor a stew pan; but a stomach, gentleman, a stomach."

An accident in 1822 literally lifted the veil which covered the human stomach and its disputed processes. Fortunately, a man of rare ability was on hand to exploit this opportunity. Dr. William Beaumont was a surgeon in the recently organized

U.S. Army Medical Corps. He was in charge of the post hospital at Fort Mackinac, on an island in northern Lake Michigan. The island was a busy center of fur trading. Trappers and voyageurs would swarm toward this post, their canoes laden with the winter's yield of fur pelts.

One such French Canadian voyageur, Alexis St. Martin, was part of the usual crowd at the trading post of the American Fur Company on June 6, 1822. Apparently by accident, someone's gun was discharged and St. Martin received the whole load, at short range, in his abdomen. The young army surgeon came on the double to aid the victim. On arrival, however, he realized that he could be of but little help. The bullets had ripped a huge wound, through which were protruding large chunks of lungs and what appeared to be ripped pieces of the stomach. Beaumont dressed the wound as best he could, cutting away some of the protruding flesh with a penknife, and had the patient carried to the shack which served as the hospital.

Had Beaumont had any colleagues on the post, he undoubtedly would have described the wound, between gulps of his supper, and would have expressed the doubts he had of St. Martin's chances of recovery. But Beaumont was all alone on this and on subsequent posts. His achievements are, therefore, all the more remarkable.

To Beaumont's amazement the sturdy youth survived the night. There followed a heroic struggle by Beaumont to save the patient; operation followed operation; no effort was spared to dress and drain the slowly healing wound. After several months the town officials refused to support the convalescent any longer; they were ready to ship him back where he came from—Canada—by open boat. Motivated by both charity and interest, Beaumont took the youth to his own home, where he continued to nurse and observe him.

For on St. Martin there was something to observe which had, undoubtedly, never before been seen by the human eye; an open window into a normally functioning human stomach! The gaping wound in the stomach never sealed. Its edges became healed, but a large hole in the upper part of the stomach remained open for the rest of St. Martin's unexpectedly long life. (He died at the age of eighty-three.)

Beaumont recognized his great opportunity, for he could "look directly into its [the stomach's] cavity and almost see the process of digestion." When Beaumont was transferred to Fort Niagara he took along his prize patient, who by then was ambulatory, and continued his studies.

Beaumont discovered that under the stimulation of entering food certain juices oozed into the stomach, and that the food disintegrated under the influence of these juices. He siphoned out some of the liquid and found that it could disintegrate food even outside St. Martin's stomach. There was no one with whom to share this exciting discovery. Beaumont continued his lonely studies patiently.

We must not visualize Beaumont's life as one of quiet ease or imagine that he could pursue his researches in a calm and serene atmosphere. Not only did he have his post duties, but, worse still, his guinea pig began to realize his own worth and became more and more demanding. St. Martin became tired of the diet which Beaumont fed him through his mouth and through the unnatural aperture, and he began to supplement it with liberal quantities of whiskey.

Finally, three years after the accident, he ran away to his Canadian backwoods. Beaumont was brokenhearted at the abrupt end of his engrossing experiments, but four years later St. Martin, a newly acquired wife, and two little St. Martins joined him again. In return for the support of the entire family Beaumont was permitted to resume his experiments.

He wanted to study the contents of these juices of the stomach, but, realizing that he was inadequately trained for the task, he tried to enlist the help of other physicians. He took St. Martin to New York but found, as he later wrote, that the doctors there "had too much personal, political, and commercial business on hand to turn their attention to physiological chemistry." He went to Yale, where he was advised to ship a bottle of St. Martin's stomach juices to the great chemist Berzelius, in Sweden. This he did, but there is no record of what became of it. If it did arrive in Sweden, it must have been in an advanced state of putrefaction, fit only for the slop jar.

Beaumont continued his studies on the increasingly unmanageable St. Martin, alone. This untrained, lone army surgeon, without equipment, without encouragement, and at his own expense, carried on his experiments testing the influence of hunger, thirst, and taste on the secretion of digestive juices. He antedated, by many years, the Russian physiologist Pavlov, who later repeated many of these experiments on dogs with artificial stomach openings.

Beaumont's discovery of the potency of the stomach's juices in disintegrating food, even outside the body, swept away all previous notions on the mechanism of digestion. The stomach is indeed not a grinding mill nor a fermenting vat, but an organ which can make a potent solution to dissolve and disintegrate the entering food. Several years later the first component of the juice to be identified was called pepsin. It is an enzyme which splits proteins into their constituent amino acids.

How does pepsin perform its work? As we saw in the previous chapter, a protein molecule is composed of amino acids. The couplings between amino acids are made by the shedding of water molecules, the amino acids being grafted together at the sites from which the water molecules are split. By boiling in strong acid the forces which keep the amino acids together are broken and, at the same time, molecules of water are inserted to patch up the shorn sites. Thus the original, intact amino acids are reconstituted. Pepsin achieves precisely the same thing; it, too, breaks bonds and adds water. How it manages to do this is the most intriguing, and the most fundamental, problem facing the biochemist, for here we come to grips with the mechanism of enzyme action—and that is ultimately the mechanism of life.

How very fundamental to life enzymes are can be realized if we consider the source of energy for all living things. The sun pours vast amounts of heat and light on our earth. The sun machine rotating in the optometrist's window converts this energy, by means of its black and silver vanes, directly into motion. No living thing can do that. Living things use the sun's energy only through the mediation of enzymes. Only green plants are able, through their enzymes, to convert the sun's energy cascading upon them into a different form of energy—chemical energy. The enzymes in the cells of all other living things can, in turn, use this stored energy for their daily needs. The following chemical reaction is the pedestal on which all of life is built:

$$\text{CARBON DIOXIDE} + \text{WATER} + \text{ENERGY} \xrightarrow{\text{ENZYMES OF PLANTS}} \text{SUGAR} + \text{OXYGEN}$$

The 20 percent of oxygen in our atmosphere is testimony of the vast extent to which this reaction has been going on in the earth's history. All of the free oxygen in the atmosphere has accumulated from this reaction. No free oxygen could be here otherwise; it is so active in combining with other elements, that when the earth was a hot, molten, "ball of fire," none of the oxygen could have escaped combination with the other elements. The reverse of the above reaction is the battery, supplying the energy for the functioning of every cell of every living thing.

$$SUGAR + OXYGEN \xrightarrow[\text{IN CELLS}]{\text{ENZYMES}} CARBON\ DIOXIDE + WATER + ENERGY$$

If we throw a pound of sugar into a burning stove it will burn, releasing exactly the amount of heat the sugar cane packed into it. The enzymes in our cells, however, perform this reaction in a controlled, slow process, releasing exactly the same amount of energy, gradually. (This happily slow reaction will be described in detail in Chapter 4.) We are lucky that the enzymes can do it in this manner, for otherwise, we would go up in a puff of smoke after a heavy meal.

Many biochemists have been attracted to the study of enzymes. Scores of different enzymes have been discovered. There are enzymes which break down proteins and others which break down fats; some enzymes have been shown to be essential for the sending of nerve impulses; the task of still other enzymes is the building up of body tissues. In every function of the body, a host of enzymes are involved.

We believe that for every chemical process which takes place in a living cell—and there must be thousands of these—there is a separate system of enzymes. They are all remarkably specific. If there is the slightest change in the material on which the enzyme functions, the so-called substrate, the enzyme becomes impotent against it.

That enzymes can function outside of the cell was shown only about sixty years ago. This milestone is the monument to the Buchner brothers who demonstrated, in 1897, that sugar can be fermented to alcohol and carbon dioxide not only by yeast cells but also by water solutions of disintegrated cells, in the complete absence of living yeast. The name enzyme originated with this discovery. Enzyme is something *en zyme*—in yeast.

How purposeful and planned the achievement of the Buchners sounds. From reading this description of their work the reader probably visualizes the brothers working feverishly in their laboratory to prove an inspired hypothesis: that yeast juice will ferment sugar just as well as the living yeast does. Like many other experimental scientists, however, they stumbled onto their discovery by sheer chance. They were trying out extracts of yeast as a medical concoction. Since they were going to feed it to patients, they could not use the usual poisonous preservative agents. So they turned to an old wives' remedy for the preservation of their extracts. It is well known to any one who makes jams or fruit preserves that a high concentration of sugar acts as a preservative. The Buchners added large amounts of sugar to their yeast extracts and the solutions began to ferment.

Actually, it was Beaumont who first saw enzymes working outside a living organism. Seventy years before the Buchners chanced upon their discovery he was digesting foods with St. Martin's cell-free stomach juice. The meaning of great accidental discoveries, such as that of the Buchners, cannot be recognized until the time, or rather the scientist's mind, is ripe for it. A great deal of knowledge had been accumulated in those seventy years: the paralyzing concept of vitalism had been abandoned; Pasteur explored the nature of fermentation. The whole scientific atmosphere was favorable to the search for a chemical and mechanical interpretation of living processes. Only when steeped in such an atmosphere could the Buchners recognize the meaning of their accidental discovery. As Pasteur put it: "Chance favors the prepared mind."

Since the time of the Buchners, biochemists have extracted
scores of different enzymes from a variety of different cells.
Every such enzyme solution contains proteins. Slowly the suspicion grew that all enzymes *are* proteins. Before 1926 chemists
were divided, however, on whether the proteins in the enzyme
solutions were really the enzymes. One school of thought, particularly among German biochemists, maintained that the proteins in the enzyme preparations were impurities and that the
enzymes were elusive, smaller molecules, present in minute
amounts. But in that year James B. Sumner, at Cornell, was
able to isolate an enzyme in a pure form, and it *was* a protein.

Sumner's achievement is so important that he was awarded
the Nobel Prize for it. He had been studying the enzyme which
breaks up urea into ammonia and carbon dioxide. This enzyme
is called urease. (The naming of enzymes is simple and uniform;
to the name of the substance on which the enzyme works, is
attached the suffix *-ase*. The scientist who first discovers the
existence of an enzyme has the privilege of naming it.) Sumner
chose a beautiful enzyme for his studies. The source, certain
species of beans, is cheap—he grew the beans himself. The material on which the enzyme acts, urea, is also cheap. Furthermore,
the enzyme produces ammonia and carbon dioxide, two of the
easiest substances to assay. This in turn makes the determina-
tion of the potency of the enzyme gratifyingly simple. The more
ammonia a given weight of enzyme can produce from urea, the
more potent it is.

Sumner isolated urease in a pure crystalline form. How does
the biochemist go about such a task? He grinds up the source
of his enzyme—in this case jack beans—with water, and ob-
tains a thin brownish soup. His obvious question is, where is
the enzyme? Is it in the soluble extract, or is it in the insoluble
bean grinds? To decide, he adds urea to a small portion of each.

The extract promptly begins to tear urea apart into ammonia and carbon dioxide; the insoluble bean debris is impotent. The enzyme is in the extract. But how much enzyme? A painstaking measurement of the ammonia that a certain volume of the extract can produce reveals the potency of the enzyme. The chemist now has his enzyme in solution but probably hundreds of other substances must be there along with it. So he begins the long, tedious task of concentrating the enzyme. He tries to throw out of solution, by means of various chemicals, either the enzyme or some of the contaminating substances. In every case both the substances thrown down and the solution remaining behind must be tested for enzyme activity. In every case the amount of ammonia formed from urea by each new preparation must be measured. As more and more inactive material is removed, the preparation becomes more and more concentrated—and smaller amounts of it will break down larger amounts of urea. After years of work and hundreds of treatments, Sumner obtained an enzyme preparation which was a crystalline protein. It is characteristic of organic substances that they do not crystallize until they are quite pure; the impurities intrude and prevent the formation of crystals. Obtaining a crystalline, pure protein which was a highly active enzyme was a great achievement. It established that at least one enzyme— urease—is a protein.

Since then, dozens of other enzymes, including pepsin, have been isolated in pure crystalline form, and every one of these, also, proved to be a protein. It took one hundred years to show that Alexis St. Martin's stomach juice owed its ability to split proteins to another protein, pepsin.

Now the biochemists know what enzymes are and what they do. But *how* do they function? How does one protein molecule —the enzyme—pry apart the constituents of another protein molecule, the substrate? Of this we know next to nothing. We are attacking the problem on several fronts, but the more we learn about enzymes the more we realize how complex is the problem and how far off the solution. Research in the field of enzymes is not unlike climbing in a strange mountain range. From a distance a peak seems near. But, as the climber proceeds, he finds hidden gullies, gigantic rock piles, and rings of smaller ridges guarding the peak, and the more he climbs, the more territory he covers, the more distant and unattainable the original peak appears.

Let us look at some of the small ridges that have been conquered in recent years. A good deal of work has been directed toward the stopping of the activity of enzymes by so-called inhibitors. We try to learn how enzymes act by learning what agents stop their action. A very small amount of cyanide stops the action of several enzymes. All of the enzymes that are inhibited by cyanide contain considerable amounts of iron. It is well known to the inorganic chemist that iron and cyanide combine into a very tightly knit compound which leaves practically no free iron in solution. Cyanide inhibits these enzymes by siphoning off their iron. (That is why cyanide is such an effective poison.) Biochemists now know that iron is essential for the action of these particular enzymes. But are we any closer to our ultimate goal? Hardly.

The study of another type of inhibition of enzymes—competitive inhibition—has not brought us much closer to the object of our quest but has yielded a whole new battery of drugs to man in his fight against bacteria: the sulfa and other drugs. (The mountaineer may never reach his peak but he may find valuable mineral deposits.)

The work of Ehrlich on the development of salvarsan is well known, but it merits continued attention because it is

the source from which flowed the sulfa drugs, penicillin, streptomycin, and other aids to therapy. Ehrlich originated chem-
otherapy. To facilitate the recognition and classification of
bacteria he subjected them to various stains and dyes. They
exhibited highly individual tendencies: some were stained by
one dye and not another, or, more interesting still, in some cases
only part of the bacterial cell was stained.

These erratically staining bacteria guided Ehrlich in his search
for new drugs. A dye stains a cell by entering into a chemical
union with its contents. Since there is such a profound demar-
cation, even within the same cell, between staining and non-
staining areas, it is entirely possible, argued Ehrlich, that there
may be some poisonous chemicals which will selectively combine with microorganisms, damage them alone, and leave the
tissue cells of the host unharmed.

He had spectacular success in exorcising the parasite which
causes syphilis, following this principle. He and his associates
kept making arsenic-containing organic molecules, almost at
random, until they hit the bull's-eye with the celebrated "magic
bullet," salvarsan. This compound kills the syphilis parasite
by poisoning some of its enzyme systems. Fortunately, in the
doses used in therapy, it is relatively innocuous to humans.

Ehrlich's brilliant discovery nurtured the hope that new
chemicals might be found which may be equally effective against
other parasites which plague us and which are unaffected by
salvarsan. The method of the search was the same as Ehrlich's
—patient testing of each compound the researcher could lay
his hands on. It was prospecting among organic compounds
for new drugs, instead of in sand for gold.

There were no guiding principles in the search. Each compound was tried on test-tube culture growths of various bacteria,
and if any showed promise by killing the bacteria they were

tested on mice or other experimental animals. For many a drug is effective in the test tube but useless in the whole animal, either because it is too poisonous or because it is rendered harmless to the bacteria by conditions in the animal. The work is slow and tedious. For example: a hundred mice might be injected with an identical dose of a virulent strain of streptococci —the little beasts which cause "strep" throats. In addition to this injection, fifty of the mice might receive the same dose of the drug which killed those streptococci in the test tube. In transferring from the test tube to the mouse the dosage of the drug is calculated by proportion. If one milligram of the drug killed the organisms in 1 cubic centimeter (about 1 gram) of broth, a mouse weighing 30 grams would get 30 milligrams of the drug. Then all that is left to do, after an impatient night, is to count the dead mice in each group. If the majority of the drug-protected mice survive while the others perish, the drug is effective.

In 1935, twenty years after Ehrlich's death, Domagk, a German physician, struck gold in a dye called Prontosil. The new drug passed all the preliminary tests with flying colors: it protected mice against the streptococci, and it was harmless to the mice. The first human saved by the drug was Domagk's own young daughter, who had come down with a severe case of streptococcal blood poisoning. At that time, the physician could only lance the wound where the organisms made the breach, and hope that the body's natural defenses would rally and exterminate the invading cocci already in the blood stream. When his daughter continued to sink, Domagk fed her large doses of Prontosil. She rallied and recovered.

This melodramatic success was the first of a series of spectacular demonstrations of the value of the new drug. Prontosil was patented by the I. G. Farben drug cartel, and the stage was set for the world monopoly of this potent drug.

"Fortunately for the world, however, Tréfouel and his colleagues in Paris soon showed that Prontosil acted by being broken up in the body with the liberation of sulfanilamide, and this simple drug, on which there were no patents, would do all that Prontosil could do." The quotation is from Sir Alexander Fleming, the discoverer of penicillin, the production of which the English scientists made available to all—in Sir Alexander's words—"without thought of patents or other restrictive measures."

Tréfouel's discovery—one of the few effective blows by a Frenchman against the Germans in that decade—gave tremendous impetus to the search for new antibacterial agents.

Sulfanilamide is a relatively simple compound; a competent sophomore in chemistry can make it. Furthermore, the organic chemist can make variants of it with ease. Onto the structure of sulfanilamide he hangs a variety of groups of atoms, and with high hopes he hands the new products over to the bacteriologist for testing.

Hundreds of altered sulfa drugs were made; some were better than the original sulfanilamide or Prontosil, others proved useless. But the search was still hit or miss. A guiding principle, an insight into the mechanism of the killing of the bacteria, was lacking.

The English scientists Paul Fildes and D. D. Woods proposed an attractive theory. It is based on a theory of Ehrlich —the lock and key theory—which that remarkable genius had proposed for the explanation of the mode of action of enzymes. The history of the lock and key theory repeats the weary pattern of the reception of new ideas. Ehrlich was ridiculed by his contemporaries, but a new generation of scientists returned with admiration to the much abused theory and used it eagerly.

Ehrlich pondered the specificity of enzymes. Why does an enzyme act on one substance and not on another? He theorized that an enzyme and the substances it can alter must fit into each other as a key fits a lock, and that the possibility of such a union determines whether the enzyme can function on a substrate.

Fildes and Woods extended the theory to its next logical step. What would happen if another substance, which simulates in appearance the normal substrate, would fit into the lock of the enzyme molecule? The key may fit, but not completely; the enzyme mechanism may jam. They pointed to the possibility that sulfanilamide may simulate in structure some substance in the diet of bacteria. Thus, sulfanilamide may crowd out the dietary essential from the lock of the bacterial enzyme.

But what is this dietary essential? It was already known that an extract of beef liver can protect bacteria against the sulfa drugs. If, to a suspension of bacteria, liver extract is added along with sulfanilamide, the drug is made impotent; the bacteria flourish.

A hunt was started to track down the substance in the liver which fortifies the bacteria against the sulfanilamide. The suspension of finely hashed liver was subjected to a variety of chemical manipulations to determine, for example: Is the factor soluble in alcohol? Can it be thrown out of solution by adding chemicals which invariably throw down proteins? In every case the material before treatment and each fraction obtained from the chemical manipulation had to be tested for its ability to overcome the toxic effects of sulfanilamide. Casting this chemical dragnet is dull and tedious work, but the hope of tracking down the active material spurs lagging spirits, and occasionally the researcher's patience is rewarded. Since unsuccessful searches are seldom presented to the lay reader, solutions of problems of this kind must sound monotonously simple. They are far from it. Sometimes, years of exhaustive—and also exhausting—intellectual and physical labor yield nothing, and the search is sadly abandoned. However, this search was fruitful. The substance was isolated in pure form and to everyone's amazement it turned out to be a well-known chemical compound, para-aminobenzoic acid. This laboratory reagent is the essential substance and, indeed, a vitamin for bacteria.

The chemical structures of para-aminobenzoic acid, nicknamed PABA, and of sulfanilamide are strikingly similar. The conjecture which launched the search was beautifully confirmed. The enzymes of the bacteria may mistake sulfanilamide for PABA. The enzymes receive sulfanilamide, but since it is not completely the same as PABA there is soon confusion in the cell. Enzyme mechanisms stall, the bacteria cannot grow and cannot reproduce. If an extra dose of PABA is given to the bacteria at the same time as the sulfanilamide, the bacteria are no longer overpowered by the sulfanilamide, and they continue to live. If the dose of sulfanilamide is again increased sufficiently, the bacteria once more will not grow. There is a definite numerical relationship between the amounts of PABA and sulfanilamide which determines whether certain bacteria can live or not. Growth of a certain species of bacteria may be stopped if, in the fluid where they live, there are 1,000 molecules of sulfanilamide to one molecule of PABA. Their ability to grow is restored if the PABA concentration is increased to a ratio of 1,000:2. The various sulfa drugs differ in the PABA-sulfa ratio which will stop bacterial growth. For those bacteria which required 1,000 molecules of sulfanilamide to prevent growth,

10 molecules of sulfathiazole will suffice. Therefore, a patient invaded by these bacteria needs to be dosed with much smaller amounts of sulfathiazole than of sulfanilamide.

Biochemists visualize the sulfa drugs as competing with PABA for the favors of some enzyme in the bacterial cell. Such inhibitors of enzymes and of bacteria are called competitive inhibitors. Studies of competitive inhibitors yield just a tiny glimpse of the enzymes' work. We conclude that the enzymes must combine with a substance like PABA. The combination is probably chemical in nature. There must be a pre-existing "lock pattern" in the enzyme into which PABA must fit, and into which sulfanilamide also fits. However, in a later step in the process, sulfanilamide must jam the machinery.

The concept of competitive inhibitions also provides us with a rational approach to the search for new drugs. We need no longer try compounds at random. There is now a guiding principle in fashioning new "magic bullets." We try to make compounds similar in chemical structure to substances which are essential for the parasites, hoping that the new compound may be a competitive inhibitor in the parasite, without injuring the host.

Our task is not simple, however, for it is an extraordinarily fortunate set of chance circumstances that renders sulfanilamide toxic to bacteria and relatively innocuous to mammals. PABA is usually absent from the bloodstream of mammals. It is but a building block for a more complex vitamin, folic acid, which mammals must acquire from their diet. The folic acid in the blood stream of mammals is, in turn, not available to most bacteria, probably because it cannot cross the barrier of the bacterial cell wall. Many bacteria are thus forced to fabricate their own folic acid from PABA, and that is their Achilles' heel, where they are vulnerable to the onslaught by the drug from Paris, sulfanilamide. At the same time the mammalian host is immune to the toxic effects of sulfanilamide, since it does not depend on its own enzyme system for the fabrication of folic acid.

Competitive inhibition is the guiding principle in the search for drugs for the therapy of cancer. Altered molecular structures have been designed to simulate every type of component ever found in the cell. Every skill of the organic chemist has been pressed into service in this vast campaign of dissimulation. Of the thousands of compounds prepared and tested several proved to be of considerable value in therapy. Unfortunately, however, the cancer cell is more than a match for the ingenuity of the biochemist. As fast as we develop new therapeutic agents the cancer cells develop mechanisms of resistance rendering the drugs impotent. Of course the ideal drug—yet to be discovered—would be one which could be selectively inhibitory only to the cancer cell as sulfanilamide is toxic only to bacteria. But in order to fashion such a drug we must know more about the enzyme systems in normal and in cancer cells. That the relative *levels* of some enzymes are different in normal and cancer cells we already know, but a privative difference is still but an object of hope and search. Until such a difference in structure or function can be found in cancer cells we must resign ourselves to the endless testing of random compounds in the search for a chemotherapeutic agent for the control of the disease.

A cell achieves its individuality and its very size by the kind and the amount of enzymes it has. Pathologists can recognize the characteristic cells of the various tissues by the presence or absence of certain topographical features and by the interaction of cells with certain dyes. The biochemist is beginning to be able to differentiate the cells of different tissues by determining the kinds of enzymes and their relative levels present. We have just started to explore the molecular mechanism which determines the levels of enzymes in a cell. As we shall see in a later chapter the potential capacity to produce an enzyme resides in the gene. But, for the maintenance of the levels of enzymes, the cell has devised ingenious regulatory mechanisms. It had to; no enzymes could be permitted to accumulate at random, otherwise the harmonious integration of cellular func-tion would be disrupted by an anarchy of competing enzymes, leading to the cell's destruction.

One of the regulatory processes is a negative feedback or cybernetic mechanism. The latter is a term coined by the mathe-matician Norbert Weiner to describe a system which is reg-ulated by its own final product. A mechanical example of a cybernetic regulator is the governor on a steam engine. This consists of two steel spheres whose rotation is actuated by the pressure of steam in the engine. The spheres are so suspended that they rise with the centrifugal force conferred on them by increasing speed. At a predetermined speed and steam pressure a sleeve attached to the spindle of the spheres rises beyond a critical height and actuates a valve which releases the steam pressure. With the reduction of pressure the speed of the engine is slackened, consequently the governor's spheres fall and the steam valve is shut. Thus, excessive speed feeds back that in-formation to the machine and self-regulation is achieved.

That such a negative feedback mechanism exists in living cells was discovered simultaneously by two different investiga-tors, D. D. Woods of Oxford University and J. Monod of the Pasteur Institute. It has been known for a long time that if bacteria which had been growing on a complex diet containing a variety of nutrients are transferred to a scanty environment containing only sugar and salts, the bacteria will resume growth only after a long lag. They were "adapting" to their new environment. What could be the mechanism of this adaptation? It was found that the bacteria were accumulating enzymes for the fabrication of some of the nutrients of which they were deprived in their meager environment. If one of these lacking nutrients, say a specific amino acid, is returned to the bacterial broth the enzyme for its fabrication disappears from the newly growing cells. In other words, a system of negative feedback communication exists within the bacterial cell which orders the halting of the making of an enzyme since there is an abundance of the end product of the enzyme's activity. The molecular mechanism of this communication system is completely obscure at the present time. Its importance, however, in biological systems is unquestioned. How does an organ achieve its shape and size? Indeed, how does the whole organism arrive at its predetermined size: why does a cat not grow up into a lion? It can be confidently assumed that just as negative feedback systems operate to control an enzyme component of a cell, the same kind of mechanisms probably control the number of cells as well.

As some enzymes are suppressed by their end products, others can be cajoled to appear by the beckoning of their substrates. Some microorganisms cannot normally use milk sugar as a source of food. Milk sugar is a complex sugar made of two smaller sugar molecules, glucose and galactose. Some bacteria thrive on either of these simple sugars, but they will not grow if their diet contains only milk sugar. They lack the enzyme needed for the cleavage of milk sugar into its components. However, after prolonged incubation the lifesaving enzyme begins to accumulate within the starving cells; the organisms begin to grow. Such "adapted" cells can thereafter always use milk sugar, provided the dietary regime is not interrupted. If there is an interruption by the feeding of the simpler sugars, the milk-sugar-cleaving enzyme disappears. One might easily conclude that here we have an example of a lifesaving teleological mechanism. However, other substances, too, can provoke enzyme formation, provided a key pattern of a few atoms present in milk sugar is also present in the substance which duplicates the action of milk sugar and the bacteria cannot use some of these enzyme-inducing substances as food. Thus the production of the milk-sugar-splitting enzyme cannot be teleologically oriented for the feeding of the bacteria; rather, it appears to be a blind response to some stimulus by a certain pattern of atoms. Just how such a pattern of atoms can evoke the synthesis of an enzyme is as obscure at present as the mechanism by which the enzyme itself functions.

While our knowledge of the intimate mechanism of enzyme action is very limited, we do have much information about the different types of enzymes with which living things are equipped. Of all the enzymes, there are none more fascinating than those that produce light. Yes, "The fireflies o'er the meadow," which "In pulses come and go," owe to an enzyme the distinction of being lifted by the poet from the obscurity of thousands of other insects. They possess an enzyme which produces light. The firefly is not the only creature gifted with an astonishing flashlight. A variety of species, from the microorganisms that light up the wake of ships at sea to large deep-sea fish which have lanterns to illuminate the otherwise eternal darkness of their hunting grounds, all use enzymes for the generation of light.

The enzyme has been named luciferase by E. Newton Harvey, the biologist to whom we owe much of our knowledge in this field. The light produced by luciferase is a cold light. The enzyme is able to generate light without wasting a good deal of energy as heat. Mechanically, we are no match for the firefly. Even fluorescent lights, which are a great improvement over the hot-filament bulbs, generate some heat.

The most wonderful of these lantern-carrying creatures are certain fish which, though unable to make light themselves, carry around millions of light-producing microorganisms in little pockets under their eyes. When they want their lights dimmed, or put out, they cover the pockets with a convenient lid.

The purpose of these devices is not only illumination; the firefly has its light on its abdomen. Some deep-sea fish have light pockets along their sides, in characteristic porthole-like patterns, which vary from species to species. It is thought that members of the opposite sex of the same species recognize each other by these patterns.

Some animals use enzyme systems for attack or self-defense. The electric eel can discharge several hundred volts, over and over, to incapacitate its prey or enemy. The largest of these creatures, some of which have been captured in the Amazon river, grow to five or six feet in length. About three fourths of the body is the electric organ which generates electricity with its remarkable enzyme system.

The electric organ of the eel is crammed with the same enzyme which is found in the nerve tissues of other animals, including man. Since the nerve impulses are relayed through electrical charges, the appearance of the same enzyme in these two entirely different tissues is not too surprising. It is one more illustration of what Emerson described as the poverty of nature. "And yet so poor is nature with all her craft, that from the beginning to the end of the universe, she has but one stuff . . . to serve up all her dreamlike variety. Compound it how she will—star, sand, fire, water, tree, man—it is still one stuff

and betrays the same properties." (Exception must be taken
to considering nature "poor with all her craft." Indeed, she is
astoundingly rich in versatility, achieving, as she does, infinite
variety with "one stuff.")

The venom of snakes is another example of assault by enzymes. There are two different types of snake venom. While
both of them attack the blood of the prey, their mode of action
is different; one causes the disintegration of red blood cells,
the other clumps the red cells together. Since intact corpuscles
are essential for the transportation of oxygen to outlying tissues,
the victim of snakebite will virtually suffocate for lack of oxygen.

Further explorations in enzymes will yield rich rewards.
Scratch the surface of almost any problem in medicine and
you expose a problem in enzymes: diabetes is a derangement
of the enzymes which tackle the release of energy from sugars;
cancer—a disease of unnaturally rapid and shapeless growth—
is probably the result of some monstrous blunder in the functioning of enzymes; aging may yet be shown to be due to the
slowing down or inhibition of some pivotal enzymes.

Whether a cut in the skin will become infected is decided,
not only by the abundance of bacteria around the wound, but
also by the abundance of antienzymes present in the blood. It
was discovered that some bacteria produce a very useful enzyme which can dissolve the protective surface coatings of their
prey. Using these enzymes as a battering ram, the bacteria can
penetrate their prey all the more easily. The enzyme had originally been named hyaluronidase but this was mercifully changed
to invasin. Invasin is inhibited or rendered ineffective by an
antienzyme, anti-invasin, which is normally present in the blood
stream but which is deficient in patients suffering from severe
bacterial infections. Too little information is available to tell

how these antienzyme shock troops of the body repel the invading enzyme; whether they are enzyme inhibitors or enzymes which gobble up other enzymes. Time and more research will tell.

At the time of the original discovery of invasin, the enzyme appeared to be of only academic interest. But research sometimes takes unexpected turns. Goethe's dictum "what is true is fruitful" has been often borne out in the growth of science. An apparently modest little discovery may swell through the work of the original discoverer, or of others, to impressive proportions, providing excitement and joy to all researchers—particularly to the one who made the initial observation. What may develop from studies of invasin and anti-invasin is impossible to predict now. A totally unforeseen practical benefit has already accrued from it: the remedy of a certain type of sterility in humans.

To make the appointed task of the sperm easier, the human sperm fluid is richly stocked with the very same enzyme the bacteria produce, invasin. In some cases of sterility it was found that the male is deficient in the production of invasin, and the sperm, unaided by this mighty trumpet, is impotent before its Jericho. The shortcoming has been effectively overcome in a number of cases by the appropriate use of preparations of the enzyme from bull testes.

So far, discussion has been restricted to what has already been accomplished in enzyme chemistry. But let us now take a peek into the enzyme chemistry of the "brave new world" of the future. The writer has a pet enzyme inhibitor, alas, yet to be discovered, which may solve a social problem that has distressed men of good will and of ill will for generations.

The extent of the pigmentation of the body is determined

by enzymes. As one of our amino acids, tyrosine, is combined with oxygen, it forms dark-colored products, the so-called melanins, which produce the color of the hair, eyes, and skin. We see the result of the complete absence of this enzyme, through an error of heredity, in albinos with snow-white hair, fair skin that cannot tan, and unpigmented pink eyes. On the other hand, we have the more abundant pigmentation of the colored races. According to our best authorities, "No pigments other than those found in the whites are encountered in the dark races," and therefore, "the colored races owe their characteristic color only to variations in the amount of melanin present"—and, in turn, to an overactivity of the enzymes which produce melanin. The difference between the blondest and the darkest of humans is only enzyme deep.

It is entirely within the realm of biochemical possibility that someday a specific inhibitor will be found which, when fed, will slow down the enzymes which produce the melanin pigments, enabling us to lighten skin pigmentation at will. Or, on the other hand, we might find an enzyme *accelerator*, as well, which will enable us to darken lighter skins. For, after all, who is to decide what is preferable? The judicious use of an enzyme inhibitor and accelerator may thus someday achieve a Utopia of color.

3 . VITAMINS

THE ENZYMES' HELPERS

WHY do we need vitamins? Why would the absence of a minute dust of white powder from a sailor's diet cripple him with scurvy? What can the vitamins do in our cells to make us their slaves?

The question of the role of vitamins in the cell will be answered against the background of an acrimonious scientific controversy which raged for almost eighty years before it was completely resolved.

Solution to a problem in science does not pop out of the head of one genius, like Minerva out of Zeus's forehead, fully developed and completely integrated into the rest of the body of scientific knowledge. Our understanding of the role of just one vitamin required the intellect and labor of four generations of scientists.

Before we get on to the controversy let us introduce the two protagonists who started the feud. In one corner is the champion, Louis Pasteur, the greatest biological scientist of the nineteenth century, perhaps of all centuries. At the age of thirty-eight, when this controversy started, he was already a famous chemist with a list of brilliant achievements to his name. About this time he was leaving the field of the chemistry of crystals and was laying the foundations of bacteriology as a science.

Later he studied the "diseases of wine" under municipal sponsorship and the diseases of silkworms under commission of the French ministry of agriculture.

From his studies, Pasteur concluded that the "diseases of wine" were produced by bacteria which contaminate the juice of the grape, and whose nefarious by-products were assaulting the French palate. He prescribed appropriate remedies: warming the unfermented juice to destroy the trespassing, undesirable organisms. Had he gone no further, Pasteur would have been universally hailed by his countrymen as a savior of his country second only to Joan of Arc. (No doubt there must have been many who would have rated delivery from the despoilers of wine even higher than delivery from the British.) But Pasteur did go further. He widened his researches and his conclusions and announced that some human diseases, too, are produced by bacteria. With this statement he came into a head-on collision with some members of the French Academy of Medicine. There ensued a series of celebrated polemics— including at least one invitation to duel—in which Pasteur showed himself not only a man of genius but also a man of iron will, ready to fight for truth as he saw it through his microscope. We get a glint of the steel in the man, in this earlier, less publicized controversy.

The fermentation of sugar into alcohol and carbon dioxide had been known for a long time, but the motivating force which induces fermentation was unknown. The German chemist Liebig had proposed the most popular theory of the time. Fermentation was supposed to be produced by the last vitalistic "vibrations" of dead biological material. (Actually, this bizarre hypothesis was proposed some hundred and fifty years earlier by G. E. Stahl, a physician to the king of Prussia.) Decades

had passed since Wöhler's synthesis of urea, but vitalism still
dominated the minds of many scientists.

Pasteur concluded from his own studies that, on the contrary, fermentation is the normal function of living yeast cells,
and that it proceeds apace with the growth of yeast cells. He
published his views in 1857 in a historic paper: "Mémoire sur
la fermentation appelée lactique."

Pasteur was not the first one to make the correlation between
living yeasts and fermentation. A compatriot of his, Cagniard
de Latour, had arrived at the same conclusion twenty-two years
earlier. Latour stated that yeasts were living organisms, "capable
of reproducing themselves by budding, and probably acting on
sugar only as a result of their growth." But even scientific truths
must have apostles to struggle for their acceptance, and the
struggle sometimes gets rough. Pasteur was exceptionally
equipped for both tasks: to see truth and to fight for it. He
proceeded to prove that fermentation did not require biological
material "vibrating" in its moribund throes. In 1860 he published a paper showing that fermentation did not require any
foreign protein. He claimed that from a solution of mineral
salts, ammonium salt, sugar, and a very small seeding of yeasts,
"the size of a pinhead," he obtained both fermentation—he
zestfully described the copious evolution of carbon dioxide—
and a lush growth of healthy yeast cells.

This was electrifying news. Here was evidence that a living
thing, a yeast cell, can grow and reproduce in a medium completely devoid of any mysterious vitalistic substance. Only sugar,
minerals, and ammonia were needed. Such a fundamental experiment was bound to be repeated.

And now, the challenger: Justus Freiherr von Liebig, the
dean of German chemists. In 1869, at the age of sixty-six, he

could look back on a life rich in achievement in organic and agricultural chemistry. Liebig announced that he could not repeat Pasteur's experiment! Furthermore, he literally insulted Pasteur by suggesting that he deceived himself by mistaking for yeast some stray molds growing in his flask. Pasteur, the greatest living expert in bacteriology and a handy man with a microscope, unable to distinguish a yeast cell from a mold filament! He replied with characteristic pungence: "I will prepare, in a mineral medium, as much yeast as Mr. Liebig can reasonably ask, provided that he pays the cost of the experiment." Furthermore, Pasteur invited Liebig to come to his laboratory so that he might repeat the experiment in Liebig's presence. Liebig was, for those days, an old man, and he died four years later, in 1873, at the age of seventy, without accepting Pasteur's defiant challenge.

The first round went to Pasteur on points.

The next important development in the controversy came in 1901, six years after Pasteur's death. Wildiers, at the University of Louvain, calmly restudied the problem of raising yeast cells in a mineral medium. He found that the crux of the problem was the size of the droplet of yeast cells used to inoculate the sterile mineral broth. Pasteur said he used a droplet the size of a pinhead. Unfortunately that was not a very exact prescription. Just as medieval philosophers are said to have debated on how many angels could stand on the point of a pin, bacteriologists began to debate the *size* of a pinhead.

Wildiers found that if the size of the inoculating droplet was very small, yeast cells did not grow in Pasteur's medium. The few that grew were sickly looking, malformed little creatures. He reluctantly took Liebig's side in the controversy. But he went further than that. He showed that with a large droplet he could repeat Pasteur's successful experiments. Furthermore

he could use a very small inoculum of live yeast cells plus a
large droplet of sterilized yeast cells, in which all the organisms
were killed, and, adding these two to the mineral broth, he
obtained lush yeast growth. He concluded that there is something other than live cells in the large inoculum which the
yeast must have for growth. He called this something bios—
from the Greek word for life. Wildiers found that bios is present in a variety of substances. A sterile extract made from meat
or from egg yolk, added to Pasteur's broth, enabled the yeast
to grow from minute inocula.

There was something present in these extracts which was
indispensable to the growth of yeast cells. Thus Wildiers demonstrated the existence of vitamins, long before the term was
coined. He tried to isolate the bios. But the current techniques
of chemistry were not up to the task.

One might expect that progress in the bios problem should
have been rapid after this. Far from it. The controversy really
began to rage in earnest. At first it was denied that there is
such a thing as bios. Experiments were brought forth showing
that yeasts do not need bios. These may very well have been
sound experiments, for we know today that different strains of
yeasts do have different nutritional requirements. Some anti-
bios crusaders held that bios was not essential; it merely over-
came the poisoning of yeast by copper. Verifying Pasteur's
famous dictum—"Nothing is so subtle as the argumentation
of a dying theory"—some went even so far as to say that yeast
grew better with extracts of meat, not because of bios, but because of some other substance. These bacteriologists, it has
been said, were "qualified to join the Last Ditch Bacon Club,
which holds that Shakespeare's plays were written, not by Shakespeare, but by some other, bearing the same name."

The bios contenders continued to wrangle; anyone who could

think of nothing more productive to do could always show that some exotic substance did or did not contain bios.

This round belonged to Liebig, the challenger.

Before we go into the final round in the controversy we must turn our attention to another area in science where big strides, which eventually led to the resolution of the bios problem, were being quietly made.

The knowledge that foods can remedy some human diseases is as old as recorded history. In some Egyptian papyri we find descriptions of the ritual by means of which the priests restored the eyesight of travelers, returned from prolonged trips in the desert. With appropriate incantations they fed to the afflicted the liver of a donkey sacrificed under suitable omens in the sky. In the Apocrypha there is the story of Tobit (Tobias, in the Vulgate), who lost his eyesight. His son Tobias was instructed to catch a monster from the sea and "anoint" his father with its liver. It is difficult to determine, at this distance, how the son might have interpreted the original term for "anoint." If he fed the "monster" liver to his father he practiced perfectly sound vitamin therapy. The lack of vitamin A in the human diet causes, at first, night blindness and, later, almost complete blindness. The richest source of this vitamin is the liver of animals, especially fish.

The cod-liver oil industry ought, perhaps, to make young Tobias its patron saint, for he was the first to practice the trade. The use of fish-liver oils in therapy in modern times was first mentioned in 1782, when the English physician Robert Darbey wrote, "an accidental circumstance discovered to us a remedy, which has been used with great success . . . but is very little known, in any country, except Lancashire. It is the cod, or ling liver oil."

How very recent is our knowledge of vitamins can be ap-

preciated from the following quotation from a leading text-
book on diet. "The chief principles in food are: Proteids
[archaic name for proteins], Carbohydrates [sugars], Fats, Salts,
Water." Not an inkling of anything else. That book was published in 1905.

Since the story of the discovery of vitamins has been often
told it needs to be summarized only briefly. A Japanese admiral,
Takaki, had a hunch that the beriberi with which sailors on
long voyages were plagued might be due to their poor diet at
sea. He was a born experimentalist, for, in 1882, he sent out a
ship well stocked with meat, barley, and fruits, and indeed, no
beriberi occurred among the crew. This was clear-cut evidence
for the relation between diet and disease, but of course there
was still no glimpse of what was lacking in the diet.

Fifteen years after Takaki's cruise a great stride was made
by a physician of a Dutch penal colony in Java, Dr. Christian
Eijkman. He noticed that hens feeding exclusively on polished
rice—the staple diet of the natives—came down with a strange
ailment. They were overcome by lassitude which progressed to
complete paralysis followed soon by death. Eijkman was able
to revive the moribund birds by feeding them the polishings
from the rice.

This was a profound discovery. In the first place, here was
an unequivocal demonstration that withholding a part of a
food from the diet can induce a disease and restoring the same
food can cure that disease. Furthermore, having an experimental animal in which a disease can be induced at will is always
a great asset. (Our inability to induce pernicious anemia in experimental animals had been a tight brake on our progress
against this disease.) When we have convenient, susceptible
experimental animals, forays can be made against a disease
from many sides: the internal changes at various stages of the

disease can be studied; a variety of possibly dangerous medications can be tried out; once a medication is won, it can be standardized.

Eijkman realized that there is a substance in the outer coats of rice which is essential for health. However, in his interpretation of his findings he went astray. Because the imprint of Pasteur's ideas and personality was so strong, every disease was attributed to pathogenic microorganisms. Eijkman thought there must be something in the rice polishings which "neutralized" the "germs" that cause beriberi.

As often happens, enough scattered information was at hand to weave a pattern of truth. Sir Frederick Gowland Hopkins, professor of biochemistry at Cambridge University, stated in 1906 that animals require in addition to the known components of their diet some unknown "minimal qualitative factors." Six years later Hopkins presented unequivocal evidence for his hypothesis. He placed a group of young rats on a diet consisting of all the known components of milk: protein, fats, sugars and salts. The weight of the rats on such an artificial diet remained stationary for twenty days. However, if he supplemented the minimal diet with 2 cubic centimeters of milk per day per rat their weight almost doubled during the same period. This was a beautifully designed, clear-cut experiment which could be repeated by anyone, anywhere in the world. The blueprint for the method of search for "vitamines" was drawn up. The catchy name was coined by Casimir Funk, one of the outstanding early workers in this field, who secured evidence that the anti-beriberi substance in rice polishings belongs to a class of organic compounds called amines. In the next three decades new vitamins, a whole alphabet of them, were discovered, but their specific task in the cell remained unknown.

At the time when Hopkins suggested the existence of "minimal qualitative factors" an important parallel discovery was made in England. The intimate relation between the two discoveries did not become apparent for thirty years. The new discovery was destined eventually to throw some light on the role of the vitamins and of bios.

The Buchner brothers were able to ferment sugar to alcohol and carbon dioxide, with cell-free enzyme extracts of yeast. (That was in 1897, forty years after Pasteur had established the nature of fermentation. Interestingly, the forty-year delay was probably Pasteur's fault. He stated that fermentation was the result of the living process of intact yeast cells. No one would design an experiment to challenge the views of so prestigious a person and so formidable a polemicist.) In 1906 two Englishmen, Sir Arthur Harden and W. J. Young, performed a challenging experiment. They took an active yeast-enzyme solution and passed it through a gelatin filter which was known to hold back very large molecules but to allow smaller molecules to go through. Neither the fraction that remained behind on the filter nor the fraction that passed through was able to ferment sugar. But, if the two fractions were pooled, the solution was as good as before in fermenting sugar solutions. For successful enzyme action, then, two factors are needed: large molecules —and we now know that these are proteins—and some smaller molecules to assist the enzymes. These were named coenzymes. But what are coenzymes? Chemists began the tedious task of concentrating solutions of coenzymes with the hope of eventually isolating them. At the same time, other chemists were working on the isolation of vitamins, for example, vitamin B_1 from rice polishings.

Before the result of these parallel searches is stated, tribute should be paid to a patient Austrian chemist who helped tremendously every other chemist who has to work with minute amounts of substances.

The year was 1910. Fritz Pregl, a professor of chemistry in Graz, was investigating the constituents of bile. Patiently he isolated a couple of hundred milligrams (28,000 milligrams make an ounce) of pure crystalline material of unknown composition. The first step he had to take to establish the composition of his precious substance was to determine the amount of carbon and hydrogen it contained. The available methods for this task required the burning of from 300 to 500 milligrams of material.

The destruction of 300 to 500 milligrams of precious substance which took years of labor to accumulate would provide but one bit of information. Nothing would be left for the dozens of other determinations and manipulations that had to be performed before the complete structure of an unknown substance could be pieced together. Pregl rebelled. He spent the rest of his life perfecting and refining methods so that determinations could be done on one or two milligrams of material. He was spectacularly successful. After the First World War, chemists came to him from all over the world to learn his methods. These disciples, in turn, spread the gospel of microchemistry, for that was the name given to this new technique. Pregl was awarded a well-deserved Nobel Prize for his work. Biochemists were now equipped with sufficiently refined tools to go digging for substances which, like vitamins, were present in their natural sources in minute amounts.[1]

We are now ready to return to the last round in the bios

[1] Still further refinements of Pregl's techniques enabled our chemists on the atomic bomb projects to master the chemistry of plutonium, from a few *thousandths of a milligram* of it. On the basis of that knowledge the vast plants for its large-scale production were built.

controversy. In 1919 a young American biochemist Roger J. Williams published his doctoral thesis, in which he stated that yeasts need "growth promoting substances" in addition to sugar, salt, and ammonia and these substances were identical with "the substances which in animal nutrition prevent beriberi." Yeasts and beasts need the same vitamin.

At the same time, Williams's older brother Robert R. Williams, a chemist at Bell Telephone Laboratories, was struggling alone, at his own expense, with the isolation of this very vitamin B_1 from rice polishings.

The next development in the history of bios was the announcement in 1936, by the German chemists F. Kögl and B. Tönnis, of the isolation of bios in pure crystalline form. They obtained, after working up 500 pounds of dried egg yolk imported from China, 1.1 milligrams of a pure substance which they named biotin, in honor of the bios problem. Their method of isolation was the usual painstaking physical and intellectual labor, stretching over several years: subjecting the dried yolk, a rich source of bios, to a variety of chemical separations; testing each fraction for its yeast-growth-promoting potency; discarding the inactive, concentrating the active fractions more and more, until the material was sufficiently purified to reward their labors by crystallizing in pure form.

Once biotin became available in pure form its structure was determined and was soon synthesized by Dr. Vincent du Vigneaud of Cornell University. Studies with the new vitamin helped to elucidate several apparently unrelated nutritional problems. Before biotin was isolated a number of researchers reported the existence of several unknown factors that behaved like vitamins. The most interesting of these is the one called vitamin H. If rats are fed raw, uncooked, egg white as the source of their protein they do not thrive. At first a skin rash appears; then they lose their hair; then they become paralyzed, and, if the diet is kept up, they die. They can be saved by cooking the raw egg white, by replacing it with another protein, or by feeding to them along with the egg white either egg yolk or beef liver.

The circumstances pointed to a deficiency disease, and a search was started for vitamin H. (The earlier letters of the alphabet had already been preempted.) At about the same time it was found that a certain microorganism whose habitat is the root of legumes needs an unknown substance in its diet or it perishes. The substance was named coenzyme R, and the search for the substance to fit the name was started. There was still another observation: certain diphtheria bacilli need a well-known compound called pimelic acid for *their* growth. After biotin was isolated and made available, it was found that vitamin H was biotin, coenzyme R was biotin, and the diphtheria bacilli used pimelic acid for the synthesis of homemade biotin.

The disease caused by raw egg white turned out to be due to the deficiency of biotin. There is a substance called avidin [2] in the raw egg white which seizes the biotin and forms with it a tightly knit combination. The biotin cannot then be absorbed from the intestine. (Avidin loses this property when it is cooked.) The disease—egg-white injury—can be produced in humans, too. Volunteers who ate a diet in which raw egg white provided 30 percent of the calories developed the characteristic skin disease in two to three weeks. "This symptom disappeared, but in the fifth week one of the group developed a mild depression which progressed to an extreme lassitude and hallucination. Two others became slightly panicky. The only striking observation in the seventh and eighth week was a marked pallor of the skin." In the ninth and tenth weeks the skin rash

[2] A contraction of avid albumin.

reappeared. The subsequent symptoms are not known for the experiment was halted and the subjects revived by adequate doses of biotin.

The depression and hallucination of one of the volunteers is very significant. This is not an isolated case of the appearance of such symptoms as a result of vitamin deficiency. Volunteers existing on diets deficient in vitamin B_1 showed similar symptoms, and the dementia of pellagra is well known. (A fuller discussion of this will be found in Chapter 12.)

Biotin, a vitamin which is essential to the health of yeasts and was discovered through research on yeasts, is apparently essential for the health of humans as well.

Biotin is effective in very low concentrations. It is one of the most potent of biological substances known. A rat needs only .03 micrograms a day. One teaspoonful of the crystals [3] would be enough to supply the daily needs of 1,000,000 rats for 100 days. Since generally in the case of drugs there is a rough relationship between the dosage of the drug and the total weight of the recipient, an approximation for humans can be made, too. A man weighing 150 pounds is about 500 times as heavy as a rat. Therefore, the approximate daily need of biotin for a man is 500 times .03, or 15 micrograms. The teaspoonful of biotin would suffice for 2,000 men for 100 days.

Biotin resolved three different nutritional problems but failed to resolve decisively the bios problem. The fact is that biotin is not bios. Something else is the bios that Wildiers described. Since biotin was isolated great strides have been made, both in the chemistry of the B vitamins and in the study of the dietary requirement of yeast. There are no less than twelve different members of the vitamin B family. Yeasts need five of these. The other seven they either do not need or they can make

[3] About 3 grams or 3,000 milligrams or 3,000,000 micrograms.

themselves. The five needed by the yeasts are the following: thiamin (B_1), inositol, biotin, pyridoxine (B_6), and pantothenic acid. Of these the last one, discovered by Roger Williams, fits the description of the original bios best. For example, the bios described by Wildiers had to be a very sturdy compound to withstand the prolonged heating customarily used in those days for sterilization. Biotin would surely have been destroyed, but not pantothenic acid. Of course, all of this is merely of historical interest, and, as in the affairs of men so in the affairs of yeasts, historical problems are not easily resolved. We do not know with certainty what strains of yeasts Wildiers used, or whether they were pure, homogeneous strains. There are hundreds of different yeast strains, and their dietary needs vary from strain to strain. Furthermore, we also know today that their dietary requirement varies with the time they are allowed to incubate. Given enough time they can make some of these vitamins themselves, among them biotin, but not pantothenic acid. But what name has been given to which vitamin is, after all, of no consequence. What matters is that an old problem has been solved, and during the course of its solution we acquired a great deal of new knowledge. And knowledge is our most valuable possession. Knowledge not only is rewarding in itself but also it often leads to new developments of practical importance. Fifty years ago the "bios problem" must have appeared ludicrous. Who but "impractical" professors would care whether yeasts need bios in their diet? But from the exploration of the dietary needs of yeasts and of other microorganisms, we learned of the existence of six new members of the vitamin B family. These vitamins improve our own nutrition, one of them is a potent drug against pernicious anemia, and finally, the ultimate in justification, millions of dollars have been made on them. The professors, alas, were not included in the last activity.

How does a vitamin function in a cell? The gross symptoms from the absence of a vitamin are obvious: animals become sick, yeast cannot grow. But what does the vitamin do within the cell to render deprivation of it so devastating? The answer came from biochemists whose interests were enzymes and coenzymes. In the decades following the discovery of the coenzyme of yeast fermentation by Harden and Young chemists were busily gathering information. It was found that the coenzymes of fermentation by yeast were present not only in yeast, but also in such varied materials as milk, animal organs, and blood. Whether the coenzyme preparation was made from extracts of yeast or extracts of frog muscle, it appeared to be identical. Yeasts and beasts need the same vitamins; yeasts and beasts have the same coenzymes.

These developments provide additional support for Darwin's theory of evolutionary ascent from some common origin. For here is biochemical evidence for the most intimate similarity between yeast and frog; they need the same vitamins; they make the same enzymes; they need the same coenzymes. It must not be inferred that frogs have evolved from present-day yeasts. The implication is that they both evolved from some common ancestral cell in which these basic enzymes and coenzymes were already present.

The coenzyme of yeast fermentation was finally isolated in pure form in 1935. A component of it proved to be niacin, which just about then was proved to be the member of the vitamin B group whose absence from the diet of animals produced pellagra. Niacin was not the only vitamin which turned out to be a component of a coenzyme. As a result mostly of the efforts of Robert Williams, vitamin B_1 was isolated in pure crystalline form and was made available for study. It turned out to be the cornerstone in the structure of a coenzyme which aids in the metabolism of pyruvic acid.

Pantothenic acid, the component of the vitamin B complex which is most likely the historical bios, is also a part of a co-enzyme—coenzyme A—which plays a pivotal role in the metabolism of fats.

Thus the real function of vitamins emerges: man and yeasts need vitamins to shape coenzymes to assist their large variety of enzymes.

How does niacin function as a coenzyme of fermentation? To form the coenzyme the vitamin is incorporated into a complex structure containing, in addition to itself, two molecules of sugar, two molecules of phosphoric acid, and another nitrogen-containing compound, adenine. The name of this complex aggregate is abbreviated as DPN. During the course of fermentation by the yeast, hydrogen atoms of sugar molecules are stripped away from the backbone of carbon atoms by an enzyme assisted by DPN. These hydrogens eventually must be combined with oxygen to form water. The hydrogens have to be transported, ferried as it were, from the sugar to those enzymes which can make the hydrogens combine with oxygen. DPN acts as the ferry for the hydrogens. It has the capacity to enter into transient combination with hydrogen atoms and thus shuttle them from one site to another. It does this in close association with the enzyme it assists. The efficiency of such a combination of enzyme and coenzyme in performing their appointed tasks is almost incredible. They can load and unload hydrogen atoms thousands of times in a minute. (The all-time speed record is held by catalase, an enzyme in our liver, a molecule of which can seize and decompose a million hydrogen peroxide molecules per minute.)

Since this book attempts to be a chronicle of ideas, not a purveyor of prescriptions, it will avoid admonition and advice on the choice of appropriate foods for obtaining the full quota of vitamins. Furthermore, we are deluged by information about vitamins these days. To cite a few of these fragments would be pointless; to do a thorough job is impossible. A book called *Vitamins in Clinical Practice* contains a thousand large pages. Moreover, there is not enough space here to relate the history of the other vitamins, although they are equally as interesting as that of biotin.

There are some aspects of our recent knowledge about vitamins, however, which have not yet been incorporated into radio commercials and which throw a light on some of the questions we have been asking.

Why do not animals, for example cattle feeding on a diet lacking in vitamins, develop the symptoms of, say, pellagra? Also, what is the reason for the large individual differences in vitamin requirements which are known to exist in different persons? The answer to both of these questions is the same: the production of vitamins by the microorganisms in the alimentary tract. The alimentary tract teems with microorganisms, most of them harmless, nonpathogenic. Not only are they harmless but, indeed, they are absolutely essential for the life of the cattle on their natural diet. The alimentary canal of the newborn calf (or child) is completely sterile. If the young calf's stomach and gut were to remain forever sterile the animal would have to be restricted to a constant diet of milk or it would perish. However, with its very first meal it acquires the founding fathers of an enormous colony of what is euphemistically called its intestinal flora. The cow, even though it has a multiple stomach, does not have the ability to digest the cellulose which makes up the largest part of its diet. The cow lacks the enzymes to split cellulose. It would starve to death with a stomachful of grass if it were not for its alimentary flora. For these microorganisms are able to convert the cellulose to smaller molecules. While doing this they keep for themselves some of the food and some of the sun energy that had been packed into the cellulose. But they "live and let live," and more than enough is left for the cow. This is a marvelous cooperative enterprise. The cow gathers the food and provides warmth for the little creatures. They pay rent with their labor, for they not only help with the cow's digestion but also provide much of its vitamin requirement. Many microorganisms can synthesize most of the vitamins for their personal needs. As they die, their cell's contents ooze out and the vitamins are absorbed by the cow.

Man, too, plays host to huge colonies of microorganisms in his alimentary canal. Many of these tiny creatures repay this kindness with their homemade vitamins. That this source can contribute a considerable portion of a man's vitamin requirement was made evident by the production of vitamin deficiencies in patients as a result of the prolonged feeding of sulfa drugs. The drugs not only killed the bacteria in the patients' tissues but also wiped out their vitamin factories by the indiscriminate destruction of the alimentary flora.

The amount of vitamin an individual must receive from his diet is dictated by two factors: his body's total requirement and the amount of vitamins his intestinal flora will make for him. Microorganisms differ tremendously in their ability to make their own vitamins; some can make almost all of them, others can make none.

There is an interesting sidelight on the possible role of intestinal flora in lengthening man's life span. At the end of the last century Metchnikoff, the Russian physiologist, was impressed by the large number of hale centenarians he found among inhabitants of the Balkan mountains. He found that sour milk was a staple in the diet of these folk, some of whom claimed to be doubling the Biblical three-score-and-ten. Their milk was soured by certain bacilli which convert milk sugar into an acid, lactic acid. Owing to Pasteur's influence, all scientists were very much preoccupied with the role of bacteria in health and disease. Metchnikoff conjectured that the large colonies of these lactic-acid bacteria in the intestinal flora may crowd out pathogenic organisms and thus promote longevity. The drinking of soured milk became a widespread fad. (Indeed, the bottles of Yogurt sold in Paris still carry the legend that its consumption is recommended by Professor Metchnikoff.) It would be nice to report now, with our newer knowledge of vitamin synthesis by intestinal flora, that there may be sound basis for enriching our intestines with lactic-acid bacteria and that the vitamins from the lactic-acid bacteria prolong the life of the Bulgarian mountaineers. Unfortunately, there is no basis for this. On the contrary, the lactic-acid-producing bacteria are among the least versatile of microorganisms in this respect. Unless their diet includes most of the vitamins these bacteria die.

With our present ignorance of the causes of aging the only prescription we have for longevity is what can be gathered from statistical studies of people who are blessed with it. Eat well, but not too much; relax; avoid infectious diseases; and above all, choose long-lived ancestors, for heredity seems to be the most important factor.

And, finally, who shall be declared the winner in the Pasteur–Liebig controversy? This frankly prejudiced referee votes for Pasteur. Remember, Pasteur's whole thesis was that fermentation is the result of the living activities of yeasts. In this he was utterly correct. Further, he stated that yeast can be grown in a mineral medium; he used sugar, ammonium, and other salts. Today we *can* grow yeast in a completely "mineral medium" of sugar and salts, plus the five vitamins. The vitamins, incidentally, are far more easily made in the laboratory or in the factory than is sugar. What of Pasteur's error in overlooking the vitamins present along with the yeast cells in his "pin head" seeding? It is said that Japanese artists purposely introduce a minor blemish in their finished paintings, for only God can make the perfect masterpiece. The "pinhead" then, was the insignificant blemish in the work of the man who was called by the great physician Sir William Osler, the "most perfect man who ever entered the Kingdom of Science."

Why was Pasteur that "most perfect man"? What are the attributes of a great scientist and why is he so rare? Pasteur's grandson, Dr. L. Pasteur Vallery-Radot addressed himself to this question recently.

Can it be surprising that the scientist of genius should be exceptional? What contradictory qualities he must possess! Besides the gift of observation, he must be endowed with imagination, so he must be a poet. He must be always ready to receive the revelation, what we have called insight; to have that readiness, he must not be narrowly specialized, his knowledge must range over widely varied fields. He must discipline himself to assiduous labor (whereas poets more characteristically wander in a dream world). He must confine himself within the bounds of rigorous experiment, requiring him to bridle his imagination. Lastly, he must have a logical mind, able to draw sound inferences and to synthesize facts observed in the course of his experimentation. Qualities so opposed to one another are very rarely united in a single individual.

Indeed they are rare. In biological sciences there has not been another one like Pasteur.

*Behold, the bush burned with fire, and
the bush was not consumed.*

EXODUS III.2

4 . SUGARS

THE FUEL OF OUR CELLS

STARCH is a large, often the major, portion of man's diet. Rice, potatoes, and flour are cheaper to produce than cheese, eggs, and meat. Therefore the majority of mankind lives mainly on those three starch-laden staples. All too many get very little even of them.

Starch is but one of several different substances which the chemist groups together as sugars. The simplest sugars are grape sugar, known technically as glucose, and fruit sugar or fructose. Both of these sugar molecules contain six carbon atoms attached in a row, festooned with six oxygen and twelve hydrogen atoms. They differ in the architectural pattern of those hydrogens and oxygens.

A molecule of each of these sugars is grafted together by the sugar-cane plant to form the cane sugar with which we are all familiar. (Fructose is sweeter than cane sugar. This accounts for the great sweetness of honey; the enzymes of the bee dismember cane sugar into its components, fructose and glucose. Saccharin, the artificial sweetening agent is not a sugar at all. It is a synthetic organic molecule which, by chance, happens to have an impact on our taste buds which induces the sensation of sweetness. Saccharin is not metabolized, therefore it has no caloric value.)

Plants can also clip together hundreds of glucose molecules to form the multi-sugars—starch and cellulose. The grafting together of the many small glucose units into the huge starch or cellulose molecule is accomplished in a simple manner: water molecules are shed, and the sugar molecules fuse at the shorn sites left by the detaching of the water.

We can but marvel at the ingenuity with which nature employs such a simple process. The fusion of smaller molecules, whether of sugars, amino acids, or fatty acids, into the appropriate large molecule is always achieved by the elimination of water.

We must remember that life started in water and continues in water. We are indissolubly wedded to water because the surface of the planet we happen to inhabit abounds in that fluid. We are "creatures of circumstance." If our earth had an average temperature a few degrees lower, the story might have been different. Water, instead of being an abundant liquid, would have been a relatively high-melting mineral, ice. (Even now much of the earth is covered with ice.) In that case, living creatures, if any could have arisen, might have had, instead of water as the major constituent of their tissues, another fluid —perhaps liquid ammonia. Chemical reactions within the bodies of such nonaqueous creatures would not involve the splitting out or addition of water. They would use ammonia as the chemical zipper in the assembly or breakdown of their large molecules.

Cellulose, which is the main component of the leaves and stems of plants, is useless to us as a food. We cannot absorb these huge molecules from our intestines; nor do we have the enzymes or the enzyme-bearing microorganisms, as do the cow and other ruminants, to break them up into smaller, usable molecules.

In whatever form the sugars are eaten, starch or cane sugar, they are broken down by the juices of the alimentary canal to simple sugars, which are then absorbed into the blood stream. All the absorbed sugars are converted to glucose; that is the only sugar found circulating in the blood. This does not, however, justify the claim of some candy advertising that dextrose (another name for glucose) is the quick energy food. Cane sugar is split in the alimentary canal so rapidly that it does not differ from glucose in its availability for a normal person.

The amount of glucose in the blood is remarkably constant; it increases in diabetes, but otherwise its level is about the same in all average healthy persons.

The body's heat is derived mostly from the "burning" of glucose; cold-blooded animals such as the frog have less sugar in their blood than we do; birds, which are warmer than we, have more. The writer could not find out whether the blood sugar of the shrew has ever been determined. This tiniest of mammals—it is smaller than a mouse—has a body temperature even higher than that of birds. Undoubtedly its blood sugar is higher, too.

If we could not store so vital a substance as glucose in our bodies, we would have to be eating incessantly to maintain a steady supply of it. That would be a precarious existence. Should we fall asleep we would never wake; for lack of its fuel our bodies would grind to a halt. On the other hand, large amounts of free glucose could not be kept in the body either; it is too readily used up. We therefore deposit glucose in a more stable, less reactive form. Scores of molecules of glucose are hooked together to form this stable reservoir called glycogen. This is the animal's version of the plant multi-sugar starch.

Glycogen is stored all over the body: there are depots of

it in the liver, in the muscles, in the kidneys. The living organism husbands and distributes its resources well. As we need
glucose, enough of the reserve glycogen is mobilized and is
broken down into independent glucose molecules to fill the
order. If there is any excess glucose in the blood, as after a
meal, it is shipped to the glycogen depots. If energy is needed,
the glucose molecule is broken down to release the sun's energy
originally packed into it by the green plant. The release of
energy is performed with an astounding series of enzyme-motivated reactions.

The earliest glimpses of these reactions were obtained, oddly
enough, from studies not of animals but of yeasts. Yeast cells,
up to a point, utilize sugars for *their* energy exactly as we do.
(They are unable to cope with alcohol, which *we* "burn" with
ease to carbon dioxide and water.)

The first step in the metabolism of glucose in the yeast cell
or in the human cell was discovered by the same chemists who
discovered coenzymes—Harden and Young. They found that
yeasts starve in the midst of an abundance of glucose unless inorganic phosphate salts are present. But when phosphates are
added to yeasts, they thrive on their glucose. Why the need
for phosphates?

Harden and Young found that the yeasts, as the first step
in fermentation, hang two molecules of phosphate on the first
and sixth carbon atoms of the glucose molecule. Later, other
chemists were able to cajole out molecules containing three
carbons with phosphate still hooked onto them. An example is
phosphopyruvic acid, a compound containing three carbon
atoms, which upon losing the phosphate becomes pyruvic acid,
the substance which accumulates in the blood of patients suffering from beriberi.

These two- and three-carbon-containing fragments of glucose

are found in yeast cells and in elephant cells. What happens to these fragments was obscure until rather recently. Their fate was revealed by the work of many biochemists and the insight of one. Sir Hans Adolph Krebs is a biochemist, trained in Germany, who found refuge in England. From isolated bits of information he pieced together the blueprint of what turned out to be the powerhouse of the cell. Pyruvic acid which contains three carbon atoms is not degraded directly to carbon dioxide. First it combines with another component of the cell which contains four carbon atoms. This seven-carbon-containing product loses carbon dioxide and becomes citric acid, which contains six carbon atoms. Citric acid goes through several enzyme-motivated convolutions, loses carbon dioxide, and becomes a five-carbon-containing product. This substance can again lose carbon dioxide and, after some enzyme motivated alterations, become the original four-carbon compound, which now can fuse with another molecule of pyruvic acid and start the cycle all over. This cyclic process has been called the Krebs cycle in honor of the man who pieced it together. It is an extraordinarily ingenious device which bestows several advantages on the cell. In the first place, the release of carbon dioxide and of energy is very gradual. Furthermore, some of the intermediate components of this cyclic system can be used for the manufacture of amino acids. If a five-carbon-containing amino acid is in short supply, the appropriate five-carbon component of the Krebs cycle is spared from further degradation and is shunted to the enzyme which can convert it to the amino acid. Finally, the Krebs cycle can "burn" not only sugars but fats as well, because fragments originating from fats can also be fed into this exquisitely controlled "fire." There may be as many as forty different enzymes cooperating to achieve this cyclic process. All of the enzymes are packed, in close proximity, within tiny compartments of the cell called the mitochondria.

These enzymes acting in patterned unison probably achieve more than the sum of their parts. I am not suggesting a new version of mystic vitalism reduced to the molecular level, but simply that the integration of the functions of a multitude of enzymes may achieve an effect which transcends the sum of the individual components. We know that proteins acquire unique properties by their very size; in turn, a constellation of such huge molecules may have functional attributes undreamed of in our present day biochemistry. The enzymes within the mitochondrion regulate themselves, take in raw materials, feed out energy and byproducts; in short, they act as a self-contained, self-regulated powerhouse for the cell.

What other enzymic constellations in other parts of the body can do we do not know. What the French philosopher Henri Bergson called the *élan vital* may well be the bubbling product of such molecular constellations.

How is the energy released from the Krebs cycle used by the cell? Before we can be qualified engineers for this most marvelous of machines we must learn the ABC of the energy of chemical reactions.

All forms of energy are interchangeable: heat can be converted into motion; motion can be converted into electricity, which, in turn, can give light or heat again. Such conversions are often very inefficient. The best steam engine loses about half of the energy of the steam as it converts it to motion.

Energy can not be destroyed; we can not circumvent its complete liberation by using different paths for its release. We can take a pound of coal and burn it in ample air to carbon dioxide. An amount of heat will be liberated. If we burn another pound of coal in a limited supply of air it will form carbon monoxide, but only about one fourth as much heat will be liberated as before. But if we now burn all of this carbon monoxide to carbon dioxide we get the rest of the original amount of heat. Whether we release the energy in one step or in several steps, the over-all amount is the same.

Now let us take inventory of the energy in glucose. Computations of energy are always based on the chemist's unit weight, or molecular weight. In the case of glucose, this is 180 grams. (One molecule of glucose weighs 180 times as much as one atom of hydrogen.) When a green plant makes glucose from carbon dioxide and water it packs energy into it. Into 180 grams of glucose—about six ounces—are packed 700 Calories[1] of energy. If we burn in a stove the six ounces of glucose we will release 700 Calories of heat.

Now, where are these 700 Calories hidden? They are used to form the bonds that hold together the six carbons, twelve hydrogens, and six oxygens of the glucose molecule. There is energy in chemical bonds. Imagine a dozen large springs from a mattress squeezed into a hat box. A good deal of energy had to be expended to squeeze the springs together before the box lid could be safely locked. If the lid is opened, the jumping springs will release the same amount of energy that was used to put them into the box. This is an analogy of sorts for the energy used to lash the atoms of carbon, hydrogen, and oxygen together to form glucose. (This bond energy has nothing to do with the energy within the nucleus of the atom; bond energy is dwarfed by the monstrous energy of the nucleus.)

The amount of energy in each chemical bond is not the same; some bonds have more energy packed into them than others.

---

[1] The Calorie is a measure of heat energy. One hundred Calories will heat one liter (about a quart) of ice-cold water to boiling.

The cell gets its warmth and its energy for work from the breaking of the bonds of glucose. Phosphates play a stellar role in the storage of the released energy in a form more convenient for the cell. The howling wind has a lot of energy. The farmer's windmill catches some of that energy and at once puts it to work pumping water. But some of that energy is also stored by the charging of batteries. The wind can not light up the farmer's house, but the battery can. The heat from a crumbling glucose molecule can not, by itself, make our legs move, but the energy in a phosphate bond can. Phosphate bonds are our batteries. They are the stored energy for life's every need.

What is the mechanism of this battery to which we owe our lives? The battery is a molecule—a molecule called adenosine triphosphate—abbreviated ATP. Onto a molecule of ATP are lashed two special phosphate groups. Ten Calories of energy are packed into each of those bonds which secure these phosphates to the ATP. These phosphate-cementing Calories are the only form of energy the cell can use for its many tasks.

Fifty such phosphate bonds are formed from the energy released by one molecule of glucose. Fifty times ten Calories are stored from the 700 contained in the six ounces of glucose.

The other 200 Calories which are not captured into phosphate bonds keep us warm. But 500 out of 700, or 70 percent, of the Calories are saved for future work. There is a loss of only 30 percent of the total energy in this transformation. The cell is thus a better engine than the best steam engine, which is only 50 percent efficient in such a conversion.

This stored phosphate-bond energy is used in an ingenious manner. Suppose the cell is in need of a substance the assembly of which requires 30 Calories of energy. Three units of ATP are alerted to act as coenzymes in the cell's assembly line; each ATP unit splits off one phosphate unit and they thus deliver the requisite 30 Calories. The three shorn ATP units in turn require replenishment of their lost energy. Glucose is mobilized from a glycogen depot and is degraded. The energy flowing from the crumbling glucose is used as a cement by the three ATP units to reattach their three phosphates. Then the glucose whittling stops; the three ATPs are ready for any new emergency.

How does phosphate-bond energy move our legs? The bones in our legs are moved by the muscles attached to them. These muscles always come in matched pairs. As one muscle contracts, its opposite relaxes; then the other one contracts, and the first one stretches. We move by such a sequence of alternate relaxing and contracting. All the work in this process is done by the contracting muscle; it moves the bone and stretches its opposite muscle. The energy for this work is provided by the phosphate-bond energy of ATP. The energy for the phosphate bond comes from glucose, and this energy in turn comes from the sun. So we are sun machines like the multivaned toy in the optometrist's window—fantastically complex sun machines, but sun machines nonetheless.

Let us lift the hood and take a look at the machinery. Muscle is made up of long fibers composed chiefly of a protein called myosin. These fibers are teeming with ATP. Dr. Albert Szent-Györgyi, a Hungarian biochemist who had already received the Nobel Prize for earlier work which included the isolation of vitamin C,[2] achieved muscular contraction in a test tube. He

---

[2] Szent-Györgyi is the perpetrator of a celebrated biochemical pun. He isolated an unknown substance which later proved to be vitamin C. He soon learned that he had on hand a sugar which had not been described previously. As the discoverer of a new substance, he had the right to name it. The names of all sugars must end with -ose, as in glucose, sucrose, and fructose. Since he had a sugar of unknown structure he submitted the name "ignose" to the British scientific journal *Nature*. The editors of that journal frowned on such frivolity and asked for a new name. "God-knows" was Szent-Györgyi's prompt reply.

managed to free the myosin threads of all their ATP. He then added to these threads, now elongated, some ATP. The long threads curled up instantly on contact with the source of energy. Contraction of myosin thread is contraction of a muscle. This demonstration is a milestone in the history of science, for in Szent-Györgyi's words: "Motion is one of the most basic biological phenomena and has always been looked upon as the index of life. Now we could produce it in a test tube with constituents of the cell."

The availability of the stored energy of ATP for movement is useful to the animal in emergencies. Sometimes the animal needs enormous amounts of energy when there is insufficient time to metabolize glucose. After a vigorous sprint to the bus one may pant for minutes before one can settle down to the calm reading of the morning paper. The panting is a forced intake of large amounts of oxygen needed for the burning of glucose to replenish the ATP used up by the exertion of hurrying to the bus.

The stretching of muscle threads lacking ATP explains the hitherto puzzling rigidity (*rigor mortis*), which sets in soon after death. The ATP present in the muscle is slowly decomposed after death, and, since the enzymes of sugar metabolism are forever stalled, the ATP is never reconstituted. Without ATP, the muscle fibers stretch and cast the corpse into the rigidity of death.

ATP has also been shown recently to be the source of energy for the light which some organisms are able to produce. It is also the source of electrical energy in nerve tissues and in the electric organs of animals which can accumulate and discharge electricity. Thus, glucose is the fuel of the cell, but the energy flowing from it is stored in this most versatile of reservoirs, ATP, which can be tapped to supply all the forms of energy a living organism can generate: heat, light, mechanical, electrical, and chemical energy. With these recent findings the biochemist of the twentieth century has completed the evidence for the mechanistic concept of life which the biologists of the nineteenth century so enthusiastically espoused.

We have not, as yet, mentioned insulin, which, as everyone knows, is essential in sugar metabolism.

How does insulin fit into that elaborate maze? An astonishing amount of work went into the winning of insulin. This work illustrates the effort needed to expose and understand the role of but one small cog in the complex machinery of the cell.

Diabetes is one of man's worst scourges. There are two different diseases which bear the same name, diabetes mellitus and diabetes insipidus. The only similarity between the two diseases is the same distressing symptom, the passing of enormous volumes of urine. The word diabetes actually means a siphon and the qualifying adjectives, mellitus and insipidus are remnants of the days in medical diagnosis when, unaided by chemistry, the hapless physician was forced to differentiate between the two diseases by the taste of the patient's urine, pronouncing the disease either mellitus (sweet) or insipidus (tasteless). Differentiating the two diseases is about all the physician could do until the early 1920s, when insulin was given to the grateful medical profession and, of course, the even more grateful patients.

Most of the early work in this field was done by physiologists, biological scientists whose interest is the function of cells and organs. They very quickly discovered one of the functions of the pancreas, an organ found just under the stomach. The pancreas produces solutions of enzymes which are poured through ducts into the small intestine, where they split proteins, fats, and sugars.

That the pancreas has other functions, too, was suggested as far back as 1686 by a physician named Johann Conrad Brunner, who thought that the pancreas was in some way involved in the utilization of fats and sugars. Two hundred years later there was complete confirmation of this hunch.

In 1889 two physiologists, Oscar Minkowski and Joseph von Mering, removed, under anesthesia, the pancreas of dogs. The dogs survived the operation, but in four to six hours began to show the characteristic symptom of diabetes mellitus; they were voiding sugar in their urine. As much as two ounces of sugar was lost by one dog in a day. At the same time the sugar in the dogs' blood increased. They became bona fide diabetics.[3]

Prior to these operations it had been known that the duct leading from the pancreas to the small intestine could be completely blocked without harming the dogs in any way. Apparently dogs could do without some of the products of the pancreas: the digestive enzymes which are poured through the ducts. The enzymes of the stomach enabled them to hobble along. The pancreas must, therefore, exert its influence on sugar utilization through some medium other than the juices pumped through the ducts into the intestine.

Complete confirmation of this dual role was provided by a brilliant operation of Minkowski. He removed the whole pancreas of dogs but he immediately grafted pieces of their pancreas under their skin. The dogs survived such operations and led a fairly normal, undiabetic life. With the pieces of pancreas inserted under their skin, there was no possibility of any en-

---

[3] The probably apocryphal story is told that it was the caretaker of the dogs who discovered the sugar in their urine. He is supposed to have noticed that swarms of bees followed the depancreatized dogs.

zymes getting into their intestines. Whatever the pancreas produced must have gone directly into the blood stream of the dogs.

Insulin was the fruit eventually harvested from these early experiments. Millions of diabetics of this and of yet unborn generations owe their painless days to these discoveries.

They were remarkably fortunate discoveries. Minkowski tried to repeat the production of diabetes in other experimental animals. He was unsuccessful with pigs, goats, ducks, and geese. Finally, with the cat, he was able to duplicate his earlier discovery on dogs. These two are the only experimental animals which develop a positive, unequivocal picture of diabetes on the removal of the pancreas.

The next great stride forward was made by a young student, Paul Langerhans, who was studying for his doctoral dissertation the structure of the pancreas. He cut thin sections of the organ and saw with his microscope two entirely different types of cells. There were grapelike bunches of cells, and among these there were small islands of cells which were different in appearance.

As early as 1893 it was suggested that it is the island cells —the islands of Langerhans as they came to be called—that produce something which is essential for the normal handling of sugars. There was confirmation of the relation between the island cells and diabetes from the examination of the pancreas of dead human diabetics. Invariably such post-mortem examinations revealed unnatural-looking, degenerated island cells.

There were immediate attempts to apply the newly found relationship between the island cells and diabetes to a possible cure of the dread disease. The spur was its wide prevalence— it afflicts about one percent of our population—and its relentless course through emaciation, muscular weakness, and final infection, to death in a few years.

The feeding of organs of healthy animals to patients with diseased organs was an ancient art and superstition. Minkowski himself was the first to try the feeding of the healthy pancreas of other animals to his depancreatized dogs. He obtained no improvement whatever. But the hope that some active principle might be extracted from a normal pancreas spurred on workers for the next thirty years.

This period was by no means fruitless. Chemists developed accurate methods to assay the sugar in the blood of animals in samples as small as a drop. Scores of physiologists plugged away at the preparation of extracts, all of which, alas, turned out to be toxic or impotent, or both, when injected into depancreatized animals.

Many were within a hair's breadth of reaching the solution, but history and the Nobel Prize Committee remember only the one who closes that final small gap. Frederick G. Banting of the University of Toronto has been acclaimed as the discoverer of insulin and has received many honors, including knighthood and the Nobel Prize. But we must remember that insulin was not the product of the "flash of genius" of one mind. Scores of scientists from Minkowski on have accumulated new knowledge, and from this pooled information arose a pattern, like a laborious jigsaw puzzle, lacking only the final fragment of information.

The futile question is sometimes asked: What if there had been no Banting? The answer is: Someone else would have extracted insulin successfully somewhat later. There were others who were following the same hypothesis and almost the same procedures as Banting.

This does not imply that there is no genius among experimental scientists. It merely means a different manifestation of genius; in the field of science, genius accomplishes what lesser minds would accomplish later. Creation in the arts is quite different. It is inconceivable that anyone but Shakespeare or Beethoven might have brought forth those very same plays and symphonies. But Newton and Leibnitz independently and almost simultaneously integrated the same mathematical abstractions into differential calculus. The artist extracts his creation almost solely from the riches of his own mind; the scientist evolves in his mind a pattern from phenomena which he and others have pried out from observations of our physical universe. Genius among scientists can be measured in years—the number of years that he is ahead of his contemporaries.

In 1920 Banting, a twenty-nine-year-old Canadian physician, read an article on surgery. It was a description of the effects of the blocking of the ducts leading from the pancreas into the small intestine. The survival of the island cells amidst the degenerating grape cells was emphasized. A brilliant idea took shape in Banting's mind. Heretofore everyone had ground together the whole pancreas in the first step of the preparation of insulin (that name had already been given to the long-sought, active agent of the islands of Langerhans). Was it not possible that the failure to extract an active preparation was due to the destruction of insulin by the enzymes of the grape cells? These grape cells were bursting with potent enzymes which gushed out on grinding. This article in surgery held the answer. Tie off the pancreas; let the grape cells wither and then try to extract insulin from the intact island cells. Lacking the facilities for carrying out this project, Banting applied to the department of physiology of the University of Toronto for help. He was joined by a young physiologist, Charles H. Best, and the two started the quest for the elusive insulin. They tied off the pancreas of several dogs, using standard surgical techniques and care on their patients. After about two months they sacrificed the dogs, removed the pancreas from each, froze these organs, and ground them up in a salt solution which simulates the salt content of the blood.

This mash was filtered free of insoluble debris and the clear solution was injected into a vein of another dog whose pancreas had been previously removed and which by then was in the advanced stages of diabetes. After the injections of these crude preparations the blood sugar of the depancreatized dog was lowered. Insulin was born!

The extraction and purification of insulin could proceed on a large scale once it was demonstrated that the enzymes of the grape cells were the enemies to be thwarted. These enzymes were known to be inhibited by acid. Therefore the whole pancreas glands of cattle were extracted in acid solutions and the extracts were purified somewhat to make them nontoxic. Such an extract was first used on a fourteen-year-old boy in the Toronto General Hospital, who was suffering from severe, hopeless diabetes. His blood sugar was immediately reduced by 25 percent. He was the first of millions to benefit from this new weapon in our all too meager arsenal against disease.

What is insulin and what does it do in the cell? We know the answer to the first question; to the second there is as yet no answer. Insulin was obtained in pure crystalline form and it turned out to be a protein. This explains why feeding whole pancreas or insulin to patients is useless; also why it was impossible to extract it until the enzymes of the grape cells were thwarted. In both cases insulin is destroyed by the protein-splitting enzymes—in the first case by the enzymes in the patient's stomach, in the second by the same enzymes oozing out of the grape cells.

Insulin is poured directly into the blood stream by the island cells. It is but one of a number of substances produced by ductless cell clumps, or glands, which regulate the activity of other cells. The fruits of these ductless glands are called hormones —a word derived from the Greek verb meaning "to rouse to activity."

The preparation of insulin has become a major industry. The pancreas glands of cattle are shipped from the slaughterhouses to pharmaceutical plants for processing. That is why a dish of sweetbreads will most likely be not sweetbreads, the culinary name for pancreas, but thymus, another gland from which the physiologist and the biochemist have not as yet been able to extract anything of value.

The potency of insulin preparations was standardized by the health organization of the League of Nations. The signal success of the international standardization of many drugs and vitamins by this organization contrasts starkly with its failures in the political field.

The exact, complete role of insulin in the utilization of sugar is not yet known. It is surprising what a small cog insulin is in the complex machinery of sugar metabolism, in which dozens of different enzymes have already been identified. But the cell's machinery is adjusted with exquisite delicacy. The slightest imbalance at any one point may pile up sugar and flood the diabetic patient with it, wreaking, like all floods, havoc in its path.

How devastating the malfunctioning of just one enzyme can be has been recently demonstrated by studies of another disease involving the faulty metabolism of sugars. Galactosemia is a rare hereditary disease whose symptoms become apparent within a few days after birth. The infant loses weight, vomits, becomes desiccated; his liver enlarges, and, in severe cases, unless the disease is recognized, the patient is lost. The afflicted cannot metabolize the sugar galactose which makes up one half of milk sugar. The other half is glucose or grape sugar. The only difference in the structure of these two sugars is the geometric pattern of the hydrogen and oxygen atoms which are attached to the fourth carbon atom in each sugar.

A normal infant is born endowed with an enzyme which attaches galactose to a coenzyme. Still another enzyme reshuffles the atoms of galactose in this combined form, converting it to glucose. Very rarely an infant is born who, through some monstrous blunder, is unable either to make the first, or attaching, enzyme or to make enough of it. One enzyme which merely helps to refurbish the atoms around one carbon atom can make the difference between life and death of an infant. However, fortunately, if diagnosis is made early and milk sugar is excluded from the diet the symptoms caused by the deficiency of this one enzyme disappear and normal growth may be resumed.

It is sobering to speculate on what our knowledge of diabetes or of insulin would be today if Minkowski and Mering and the physiologists and biochemists who followed them had been forbidden to use cats and dogs, or, for that matter, any experimental animals. Such speculation is not idle, for there have always been a small number of people who have banded together into antivivisection societies whose main function is agitation for the proscription of all animal experimentation.

It is one of the strengths and beauties of a free society that little bands of people like this, can bob up and down on its outer fringes unmolested, while the main stream of society flows on majestically, unconcerned by the little bands which are frantically pouring out a variety of pamphlets, calling on the main stream to be diverted or dammed.

Could these antivivisectionists have their way, however, not only would all research cease, but a great many drugs now available could not be used, for, to insure their safety, they must be tested on experimental animals. Close to sixty people were killed a few years ago by a new drug preparation which had not been tested on animals. These corpses, robbed of their spark of life by an "Elixir of Sulfanilamide," are testimony to the absolute necessity of testing a new preparation on animals. Sulfanilamide is quite insoluble in water and therefore cannot be dispensed in a convenient aqueous solution. An obscure pharmaceutical manufacturer tried other liquids as a solvent. He hit on ethylene glycol, and blithely sold such solutions of sulfanilamide to be taken by the spoonful. Unfortunately, ethylene glycol is converted in the body to oxalic acid, a potent poison.

Had the preparation been tested on just one dog first, those people would not have gone to their early graves. What human wealth might have been saved by the life of just one dog. Just one dog, spared from the hundreds of thousands destroyed annually at dog pounds, to test the "Elixir of Sulfanilamide." In the year of this disaster, in New York city alone, 55,000 stray dogs and 150,000 cats were destroyed at the pounds by asphyxiation.

5. ISOTOPES

TRACERS FOR EXPLORING THE CELLS

THAT FOODS are "burned" in our bodies to carbon dioxide has been known since 1789. We are indebted for the knowledge to the great French chemist Antoine Laurent Lavoisier, who founded modern chemistry and, in a limited sense, biochemistry as well. He boldly declared: "La vie est une fonction chimique."

Lavoisier elevated chemistry to a science by eschewing speculation on the nature of chemical reaction in favor of observation and measurement. He was not the first to apply quantitative methods to the study of chemical or biological systems. But he was one of the first to recognize the imperative necessity of controlling the complete environment of a reaction before a measurement can have any meaning. For example, the Flemish chemist and mystic J. B. van Helmont (1577–1644) set out to determine what a willow tree is made of. He planted a small tree in a previously weighed quantity of dry earth, watered the plant for five years and noted at the end of this time that the willow tree gained 164 pounds and the earth lost only 2 ounces. From this leisurely experiment Van Helmont concluded that all the increase in weight in the tree came from water. It is ironic that not only was Van Helmont aware of the existence of gases (he coined the term from the Greek, *chaos*); he was the discoverer of the very gas he neglected to take into account, carbon dioxide.

A view on the nature of combustion long held prior to Lavoisier was also based on experiments with a similar deficiency. G. E. Stahl (1660–1734), who first ascribed fermentation to be the product of the vitalistic vibrations of dead biological material, also provided an imaginative explanation for the mechanism of combustion. When a substance burns, said Stahl, "phlogiston" escapes from it. The phlogiston theory suffered a setback when it was shown that some substances *gain* in weight on combustion. A quick recovery was made, however. Phlogiston, it was decided, was a versatile entity possessing either a positive or a negative weight.

Lavoisier laid the phlogiston theory to rest when he showed with well-designed experiments that combustion is the result of the combination of substances with the "salubrious and respirable portion of the atmosphere." After exploring the nature of combustion of inanimate objects he undertook a similar study of "combustion" in living organisms. With astonishing experimental skill and judgment he showed that the process was the same in both the living and nonliving worlds. A guinea pig and a burning candle both produce carbon dioxide. Furthermore, he measured the amount of heat liberated by the candle and the animal and showed that the heat in each case was proportional to the amount of carbon dioxide produced. (He overlooked the heat liberated by the production of water.)

This genius was ordered to trial by the National Convention and was guillotined in 1794. He was denounced by Marat, whose own ambitions as a scientist he had thwarted. Marat had published his own theory on the nature of combustion, which was essentially a rehash of some stale alchemistic concepts. He denied that oxygen has a role in the process. Lavoisier had proved the views of Marat, the would-be scientist, wrong; but Marat, the politician and rabble rouser, unfortunately had the last word. After the revolution began, Lavoisier became a constant target of vituperative attacks in Marat's newspaper. His crime was that he had been a farmer-general of France. Although during his tenure of office this notorious tax-collecting agency had undergone many reforms, Lavoisier was nevertheless accused and convicted of guilt by association with the once corrupt agency. Testimonials to his great service as a scientist to France and to the cause of the Revolution— he was in charge of standardizing weights and measures—were of no avail. It is reputed that Lavoisier requested a two weeks' stay of sentence so that he might finish some experiments on respiration, but the presiding judge, one Pierre Coffinhal, is said to have replied: "The Republic has no need of scientists; justice must follow its course."

For a long time after Lavoisier's death, life was compared to a burning candle. However, this analogy could not be maintained indefinitely, for obviously a living organism, unlike a candle, is not consumed in its own flame. After the combustion engine was invented, the living organism was compared to that. (Man often belittles the grandeur of life's machinery by comparing it to one of his own handiworks. These days it is the vogue to compare the brain to an electronic computing machine.)

For a long time food was thought to be merely a fuel for the machine of the body. That the quality of the "fuel" was as important as its quantity did not necessarily conflict with the image of the body as a combustion engine.

However, it slowly became apparent that the living machine

is a most unusual one: the parts of the machinery themselves seem to be burning as well as the fuel. The American biochemist Otto Folin, who was appointed, in 1906, the first professor of biological chemistry at Harvard, demonstrated that tissues, too, are broken down independently of the breakdown of foods. He found the excretion of certain waste products from our bodies to be unaffected by the amount or kind of food. He attributed to the metabolism of the tissues themselves these constant waste products.

To what extent, if any, the components of the diet interacted with the tissues could not be determined. For, once the food enters the blood stream—after digestion in the alimentary canal—it disappears. It becomes hopelessly intermingled with those many substances similar to it that are already present in the tissues. Until recently, it was impossible to distinguish such molecules, which were already components of the tissues, from molecules only recently absorbed from the digestive tract.

For example, what happens to a pat of butter we eat? Before we can trace the path of that butter in the body we must first become acquainted with a little of the chemistry of butter. Butter is a fat. A typical fat molecule is built from three molecules of fatty acid and one of glycerol (the glycerin of the pharmacist). Fatty acids are made principally of carbon and hydrogen. The carbon atoms are strung together like beads with two hydrogen atoms linked to each carbon bead. The last carbon of the chain has no hydrogens. Instead, two oxygen atoms are attached to it. This aggregate of three atoms is the acidic group. The chains of carbon vary in length; the shortest is acetic acid (present in vinegar), made of two carbons; the next longer one (present in fats), contains four carbon atoms, the next six, and so they proceed with two carbon increments, up to twenty-four. There is always an even number of carbons, never an odd number, in any natural fatty acid. A natural fat such as butter contains a variety of fatty acids, short and long.

By means of their three acidic groups, three fatty acids are attached to a molecule of glycerol. The coupling between fatty acids and glycerol takes place by the shedding of water. The product made by the coupling of three fatty acids and of glycerol is the fat molecule.

If we cook a fat with hot alkali, we dismember the fat molecule and convert it back into its components: the fatty acids and glycerol. The fatty acids combine with the alkali to form a soap. This operation is the basis of the tremendous soap industry. Soapmaking has been practiced essentially the same way for thousands of years. Any available fat—beef tallow or a vegetable oil—has been cooked with whatever alkali was available. The pioneers in the American wilderness leached out the alkaline ashes of burnt wood; today we use lye produced with electrical energy from table salt. Fundamentally, all soaps are the same. They differ in color and odor, and in the amount of air, salt, and water put into them.

Butter is broken down by enzymes, mostly in our small intestine, into fatty acids and glycerol. The enzymes do, with cool efficiency, what the alkali does in the hot cauldron. The enzyme-produced fragments are absorbed from the intestine and disappear. Where they go and what they do was a complete mystery until twenty-five years ago. Of course we knew that eventually they must be crumbled down to carbon dioxide and water, but nothing more was certain.

Are they disintegrated immediately, or do they linger in the body for a while? If they linger, where are they? Do they go to the liver, or do they go to the blobs of fat depots which form all too readily around our waistlines? Do they fall apart into

carbon dioxide and water at once, or do they disintegrate gradually? Do the different fatty acids have the same nutritional
value?

These are the questions which challenged the biochemists'
imaginations. They appeared to be questions doomed to dangle
before us without answers. For once the fragments of the fat
molecule are absorbed from the intestine they become hopelessly intermingled with the pool of other fat molecules already
present in the body and become indistinguishable from them.

There have been attempts to label a fat, to hang a bell on
it, in order to chart its wandering throughout the body. In one
such attempt some of the hydrogens in a fat were replaced by
entirely different atoms of another element, bromine. The
strategy of this experiment was simple. The presence of bromine is very easy to detect and normally the amount of bromine
in tissues is very, very small. If after eating the bromine-labeled
fat a rat's liver should contain a large amount of bromine, that
would be an indication that the fat entered the liver from the
intestine.

While the scheme sounds simple and effective on the surface, actually, labeling a fat with bromine yielded no worth-
while information. Such a drastically altered fat, in which
bromine atoms had replaced hydrogen, is an unphysiological
substance. The enzymes of the body, which are notoriously
fastidious in the choice of substances on which they act, may
have nothing to do with such unnatural substances. A label
which could go unnoticed by enzymes was needed. Such a label
for the fat molecule was provided, not by biochemists, but by
atomic physicists. With the aid of this label we not only tracked
down the fat molecule but also learned so much that we changed
our whole concept of the mechanism of the cell.

Since isotopes have been seared into our minds by the heat

released over Hiroshima and Nagasaki only a brief summary of them need be presented here.

The atoms which compose an element are not uniform in weight. The vast majority of hydrogen atoms have the same weight. But there is one atom in every 7,000 which weighs twice as much as one of its more abundant lightweight brothers.

In the case of nitrogen most of the atoms are 14 times as heavy as the common hydrogen atom. (Their atomic weight is 14.) But one nitrogen atom in 270 is 15 times as heavy. (Its atomic weight is 15.) The atomic brothers of different weights are called isotopes. They are identical twins in all respects except their weight.

A light hydrogen atom is composed of a nucleus of a single, positively charged speck of matter, a proton. Rotating rapidly around this proton is a still smaller speck—a negatively charged electron. A heavy hydrogen atom also has only one satellite electron, but its nucleus is different: it contains, in addition to a proton, a neutron—a particle almost equal in weight to a proton, but without a charge on it. Since isotopes differ only in their nuclei, which are not involved in ordinary chemical reactions, their chemical behavior is identical. We know that the heavy and light isotopes enter into ordinary chemical reactions exactly the same way. But what about chemical reactions within the cell? Do the ultrafastidious enzymes differentiate between isotopes? There is a way to decide this question. If the enzymes which build up the body tissues discriminate between the isotopes, one or the other of two isotopes should be more concentrated in the cell than in the inorganic world.

Let us examine the nitrogen atoms which are built into the proteins of our tissues. Let us choose an expendable tissue such as hair for our studies. With appropriate chemical manipulations we can obtain billions of atoms of pure nitrogen from a few strands of hair. Where did these atoms of nitrogen come from? Originally they came from the atmosphere. But before they entered our body, the nitrogen atoms had sojourned in the bodies of many different plants and animals and therefore had participated in a multitude of enzyme-motivated reactions.

Eighty percent of the atmosphere is nitrogen containing the two isotopic varieties, nitrogen 14 and nitrogen 15. (Both of these varieties of nitrogen are stable; unlike the uranium isotopes, they are not radioactive.) Plants and animals are unable to use nitrogen gas from the atmosphere directly. They lack the enzymes for this task. Certain bacteria which grow on the roots of legumes have such enzymes to incorporate the nitrogen from the atmosphere into their cells. From such bacteria a clover might absorb the nitrogen atoms destined for the hair of a human being. If the clover is eaten, the nitrogen, now incorporated into amino acids, would pass into an animal. The nitrogen might be returned to the soil either by excretion or upon the death of the animal. From the soil another plant might absorb the nitrogen. (In this so-called "fixed" form, the nitrogen can be utilized by plants without the aid of nitrogen-fixing bacteria.) Thus, the nitrogen atoms which are now incorporated in human hair may have sojourned in the cells of hundreds of different plants and animals; but in all the multitude of chemical reactions inside a vast variety of different cells the nitrogen isotope was neither concentrated nor diluted. For the proportion of heavy isotope in the nitrogen obtained from hair is exactly the same as that of the heavy nitrogen isotope in the atmosphere.

The enzymes of the cell cannot tell the isotopes apart. But the physical chemist, with his instruments, can. If the isotope is an unstable one the task is easy. The radioactivity resulting from the spontaneous disintegration of an unstable isotope is very conveniently measured with the instrument designed to be sensitive to such disintegrations, the Geiger counter. There is a constant low level of radioactivity in every living cell. However, if we introduce a radioactive isotope of an element into the cell, the radioactivity increases in proportion to the amount of the isotope introduced.

The stable isotopes entail more work for the researcher. Since they do not send off telltale signals, their presence can be known only by measuring the masses of the isotopes. This, unfortunately, is a much more tedious procedure requiring great experimental skill.

The first scientist to use an isotopic tracer in biology was George Hevesy, a young Hungarian physicist working in England. In 1923 he immersed the roots of bean plants in solutions containing a radioactive isotope of lead. He traced the ascent of the lead into the stem and the leaves of the plant by simply measuring their radioactivity.[1] Of course, since there is no lead, only carbon, hydrogen, and oxygen, in a fat, Hevesy's studies were of no help in the study of the wanderings of fat in the animal. But his success pointed the way to the use of similar "tracers" should they become available.

The labeling of a fat had to wait more than ten years until

---

[1] According to Dr. Hevesy, his first "tracer" experiments were designed for a somewhat less exalted purpose. Since he was suffering from the endemic affliction of young scientists of that period, impecuniousness, he was lodging in a very modest boarding house. He observed that roast beef was invariably followed by hash on the weekly menu. He had strong suspicions that even the meager leavings of the roast beef from the plates of the boarders greeted them again the following day disguised as hash. Therefore he once sprinkled a trace of radioactivity on his leavings from a meal and next day he surreptitiously carried off a sample of hash to the laboratory. His instruments confirmed his suspicion: the hash was radioactive.

Harold C. Urey separated the heavy isotope of hydrogen—deuterium—from the light ones.

The pioneers in the use of deuterium in the study of the over-all fate of foodstuffs—or their metabolism—were Rudolph Schoenheimer and David Rittenberg of Columbia University. As the isotopes became more readily available other research teams were formed, until today there is hardly a biochemical laboratory which does not use isotopes as a tool.

To carry out an investigation using isotopes was no easy task. First of all the test compound enriched with isotope had to be prepared. A fat was made which contained not an ordinary fatty acid, but a fatty acid in which an abnormally high number of the atoms were replaced by the heavy isotope of hydrogen—deuterium. This deuterium-containing fat was fed to rats kept in suitable cages so that their urine and feces could be easily collected.

To the amazement of every biochemist and physiologist only a small fraction of the deuterium appeared in the excreta of the animals the next day. Apparently the dietary fat is not burnt up immediately in the body.

Where was the newly eaten fat? Where was the deuterium stored?

Rats were killed one, two, three, and four days after the feeding of the original deuterium-labeled fat. The various organs—the liver, the brain, as well as blobs of abdominal fat were separately cooked with alkali. The released fatty acids from each were isolated and the deuterium in them was determined. Most of the deuterium, therefore most of the original fat, was in the fat depots.

The first intelligence gained from our isotopic detection then, is that most of the fat from the diet enters the fatty depots first. How long does the fat stay in the fatty depots? The longer the rat lived after the original isotopic meal (they were on a normal rat diet after that), the less deuterium remained in its body. In three days one half of the deuterium, therefore one half of the newly acquired fat, had been used up from the fat depots. In other words, the fat which is incorporated into the body after a meal is used up only slowly. But the animals were not losing weight, therefore as the depot fat was used up, it must have been replaced from more recent meals.

This new intelligence revolutionized our whole concept of life's economy. Previously it had been thought that entering food was immediately burned up. Only the excess over the body's daily requirement was believed to be stored in depots. These depots, in turn, were believed to be inactive reservoirs tapped only on lean days. The living organism had been visualized as a combustion engine, receiving its fuel—the food—and converting it into energy and waste products without any alteration in the structure of the engine.

But the isotopes told a different story. The information obtained from studies of the fats, later confirmed with other foods, proved that the body is in a constant state of flux. Its tissues are being built up and broken down simultaneously. The molecules which compose our bodies today will be gone in a few days and replaced by new ones from our foods.

The body as a combustion engine is approximated by the train on which the Marx brothers were once escaping from one of their dire predicaments. For want of fuel in the coal car they tore up the coaches, feeding the wooden planks into the engine. Had they been repairing the coaches at the same time from fresh supplies of lumber they would have almost simulated the engine of the body.

The late Dr. Schoenheimer offered a regiment as an analogy for a living organism: "A body of this type resembles a living adult organism in more than one respect. Its size fluctuates only within narrow limits, and it has a well-defined, highly organized structure. On the other hand, the individuals of which it is composed are continually changing. Men join up, are transferred from post to post, are promoted or broken, and ultimately leave after varying lengths of service. The incoming and outgoing streams of men are numerically equal, but they differ in composition. The recruits may be likened to the diet; the retirement and death correspond to excretion." He added, however, that this analogy is "necessarily imperfect."

It is impossible to evoke a perfect analogy to a living organism. It is trite but true that the only analogy to a living organism is another living organism.

Another bounty from the study of fats with the aid of isotopes is the solution of an old riddle in the metabolism of fats. It had been known for a long time that animals can convert sugar or starch into fats. An animal receiving but a small amount of fat in its diet along with large amounts of starch produces huge fatty depots far in excess of the total amount of fat eaten. The fattening of cattle on a diet of corn, which is high in starch and low in fat content, is a practical evidence of this transformation. The corn starch is converted by the cow's cellular alchemy into the characteristic firm streaks of fat in a good cut of beef. A goose must suffer a fatty enlargement of its liver to produce a *paté de foie gras* to a gourmet's taste. It is forcibly fed huge amounts of starchy meals; corn is shoved mercilessly down its unwilling throat. It produces so much fat that its liver becomes gross and gorged with fat.

To see whether animals can really make all of their fats, an interesting experiment was carried out by a husband and wife biochemical team, George O. and Mildred M. Burr. Rats were kept on a diet completely devoid of all fats. Such a diet contains about 25 percent of casein from which all fats are removed with ether, which dissolves the fat but not the protein. The diet also contains over 70 percent of cane sugar. Commercial cane sugar, which is so pure it contains nothing else, is used. The rest of the diet consists of a salt mixture and all the vitamins.

At first rats do quite well on such a diet. But after a while it gradually becomes apparent that something is wrong; the rats do not put on their daily quota of weight. In about seventy days the rats are a sickly looking lot. Their tails are scaly, ridged, and fragile; pieces of the tails crack off. Their hair is full of dandruff and falls out in clumps. That their internal organs are also damaged is obvious from their bloody urine; the kidneys must be cracking too. If the rats' diet is not changed, these kidney lesions will kill the animals. But if before the rats reach a moribund state a few drops of fat—lard or a vegetable oil— are given to them daily, they miraculously recover.

What is there in a fat which protects the rats against this loathsome disease? Besides their varying lengths, fatty acids differ in another respect. They do not all contain their full complement of hydrogen atoms. In a normal, or saturated, fatty acid there are two hydrogens for each carbon atom. However, in some fatty acids there are carbon atoms with only one hydrogen allocated to them. These are the so-called unsaturated fatty acids. Usually a fat contains a mixture of saturated and unsaturated fatty acids, the liquid fats being richer in the unsaturated fatty acids.

The various components of a fat were fed to the rats made sick by the lack of fats in their diet. The charm which protected them was not glycerol; it was not the saturated fatty acids; it was an unsaturated fatty acid called linoleic acid.

If minute amounts of this doubly unsaturated fatty acid are fed along with the fat-free diet, the rats lead a normal healthy existence. Why must rats receive this fatty acid and not the others in their diet? Isotopes wielded by Doctors Rittenberg and Schoenheimer gave the answer.

A group of mice on a normal diet received injections of water in which the hydrogen atoms were replaced by their heavy twins deuterium atoms.[2] (Such water has a greater density than ordinary water and is therefore called heavy water. It looks and tastes like ordinary water; only the sensitive instruments of the physical chemist can tell the two apart.) A few days after the injection the mice were killed and their various fatty acids separated.

The saturated fatty acids contained large amounts of deuterium. How did deuterium get into these fatty acids? It could be there only if the mice made these fatty acids, using hydrogen from their body water to string onto the carbon skeleton. After the injection of the heavy water their body water contained not ordinary water but deuterium-enriched heavy water. Since the enzymes do not discriminate between the two isotopes, both ordinary hydrogen and deuterium were strung onto the carbon skeleton of the saturated fatty acids.

The charm, linoleic acid, which wards off the disease produced by the fat-free diet, was also isolated from these mice. It contained no deuterium at all. This is absolute proof that the mice could not make this compound; if they could, there would have been deuterium in it.

The isotopes taught us a new profound lesson. The animal's cells need linoleic acid for some unknown purpose.[3] Since they

[2] Heavy water was rare and expensive. Only one fifth as much heavy water is needed for an experiment with a mouse as with a rat.
[3] Fats from animal origin are much lower in unsaturated fatty acids than are fats produced by plants. Moreover, we destroy some of the unsatura-

cannot make linoleic acid they depend on their diet to supply this precious, essential substance. The animal body must have a huge variety of different substances in its cells for their healthy, smooth functioning. The cells can make most of these. Those they cannot make must come from their diet or they perish. Hence the need for vitamins and, as we shall see later, for essential amino acids and a few other miscellaneous substances such as linoleic acid. (Some biochemists consider linoleic acid a vitamin.)

Some species of animals are not slaves to all of these essential substances. The rat, for example, never comes down with scurvy. It needs no vitamin C from its diet; it can make its own. Plants can make all of their amino acids and vitamins from salts, water, and carbon dioxide. Animals are, in a way, parasites living on the plants. Some microorganisms are parasites, too. We differ only in the degree of parasitism. The yeast can do very nicely on sugar and five members of the vitamin B family. The red bread mold, *Neurospora*, needs only sugar and one vitamin, biotin.

How did animals become so nutritionally dependent? How did they forget the know-how of vitamin making? *Neurospora*, the red bread mold, answered these questions for us. The story of *Neurospora* and what has been learned from it will be related in Chapter 10.

While animals have qualitative shortcomings in their manufacturing abilities, work with isotopes revealed that prodigious synthesizing activities take place in animals as well as in plants.

---

tion of plant fats by artificial hydrogenation. It has been suggested that the American diet which is rich in animal fats and poor in unadulterated vegetable fat is a contributory factor to the aberrant metabolism of cholesterol which may lead to arteriosclerosis. The evidence for this correlation is so far only circumstantial. Research in this area is handicapped by the difficulty in producing arteriosclerosis in experimental animals.

Plants are able to fashion a large variety of complex products from carbon dioxide—a molecule containing but one carbon atom. Using the energy of the sun, plants are able to lash together several molecules of carbon dioxide to form elaborate structures. Animals lack the ability to use the energy of the sun directly. But with the stored energy of the sun in the form of carbohydrates, or more specifically of ATP, animals, too, can synthesize complex molecules from smaller carbon fragments including carbon dioxide.

That carbon dioxide can be used for the shaping of larger molecules by other than plant cells was first shown by G. Harland Wood and C. H. Werkman of the University of Iowa. This team composed of a biochemist and a bacteriologist showed that bacterial cells, though lacking chlorophyll, can attach carbon dioxide to a three-carbon-containing molecule to form a larger molecule of four carbon atoms. Their discovery elevated carbon dioxide from the role that had been ascribed to it earlier—merely a waste product in bacterial and animal metabolism. The demonstration of the new synthetic role of carbon dioxide in bacterial cells was made by chemical means. Wood and Werkman showed by classical analytical methods the accumulation of the four-carbon-containing molecule.

The unequivocal demonstration of a similar function of carbon dioxide in animal cells had to await isotopic tracers, for it is much more difficult to show the formation of an intermediate product in animals. The complex metabolism of animal tissues does not permit the abnormal accumulation of any such intermediate. A research team of a biochemist and a physicist, E. A. Evans and L. A. Slotin [4] of the University of

---

[4] Dr. Slotin, a Canadian physicist, was killed in a tragic accident while working on the atom bomb project. At the end of the war, just before he was to return to the University of Chicago, while he was instructing his successors, a mechanical failure caused the release of intense radiation. Dr.

Chicago, proved that animal cells can duplicate the feat of a plant cell; they too can use carbon dioxide as a building block for the assembly of larger molecules. Evans and Slotin used carbon dioxide in which the carbon atoms were a radioactive isotope. By means of a Geiger counter they were able to show that the radioactive carbon dioxide was incorporated into sugar molecules within the animal cells, for the sugar they extracted after the administration of isotopic carbon dioxide was radioactive. Such a demonstration would have been impossible without an isotopic tracer. Ordinary chemical analysis could show only a *net* increase or decrease in the sugar content of the tissues. But since the components of tissues are in a state of flux with both synthesis and degradation going on at the same time, the formation of new molecules from carbon dioxide would not necessarily result in a net increase. A corresponding number of molecules might have been degraded in the dynamic tug of war between breakdown and synthesis within the tissues.

That very extensive synthesis takes place in animal cells using two-carbon-containing fragments as the building blocks has also been shown with the aid of isotopes. Cholesterol, a complex molecule containing twenty-seven atoms of carbon, forty-six atoms of hydrogen, and one of oxygen, is made from several molecules of the two-carbon-containing fatty acid, acetic acid. (If acetic acid which contains isotopic carbon is fed to an animal its cholesterol will contain large amounts of the isotopic carbon.)

Very little is known about the metabolism of cholesterol. This complex molecule is present in every cell of the animal body; it is particularly abundant in nerve tissue. In the degenerative disease arteriosclerosis, there is a rise in the choles-

---

Slotin was able to remedy the failure in time to save his colleagues, who were at a distance from the source of the energy, but he succumbed within a few days to the effects of the lethal radiation.

terol content of the blood and its deposition in the walls of the blood vessels causes their "hardening." The vessels lose their elasticity. Although cholesterol had been known since 1775, its chemical structure was decoded only about thirty years ago. It was found that the kernel of cholesterol structure appears in a variety of hormones—the sex hormones and the hormones of the adrenal cortex, including cortisone. It had been conjectured that one of the functions of cholesterol might be to serve as the raw material for the production of some of these hormones by the appropriate glands. That the conjecture was valid was proved by a biochemist who wielded his isotopic tools with ingenuity. Dr. Konrad Bloch, at Columbia University, prepared, by chemical methods, cholesterol with large amounts of heavy hydrogen in the molecule. If, after feeding such labeled cholesterol, some of the hormones of an experimental animal contained heavy hydrogen it would clearly demonstrate that the precursor of the hormone was cholesterol. The choice of the appropriate experimental animal was crucial for the success of the experiment. Since they are so potent, hormones are made only in minute quantities in the animal glands. Feeding the labeled cholesterol to a small experimental animal and then attempting to isolate a hormone from one of its glands would have been futile: the amount of hormone present is too small for successful chemical manipulation. Feeding the precious cholesterol to a large animal such as a cow or a horse would have been prohibitive. However, it was known that some of the sex hormones are normally excreted through the urine, but only in minute quantities (12,000 quarts of urine yield 10 milligrams of the male sex hormone). But, it was also known that pregnant animals excrete one of the sex hormones—pregnanediol—in somewhat larger quantities. The labeled cholesterol was fed to a pregnant woman and from her urine the pregnanediol was isolated. It contained considerable amounts of heavy hydrogen. Therefore the precursor of that hormone must have been cholesterol. In turn, the precursor of cholesterol is acetic acid. Thus, the acetic acid in a salad dressing eaten by a pregnant woman may be incorporated into a hormone molecule by the virtuoso synthetic processes of the human body.

An interesting application of our knowledge of isotopes in biological material is the dating of ancient remnants of life by a method devised by Dr. Willard Libby of the University of Chicago. The cosmic radiation which is constantly bombarding us converts a very small but constant amount of nitrogen in our atmosphere into radioactive carbon (C^{14}). This C^{14} is inhaled by plants as carbon dioxide and is thus incorporated into every living organism; C^{14} decomposes at a slow and steady rate so that one half of it disintegrates in 5,900 years. During its decomposition C^{14} gives off beta particles which can be determined in a Geiger counter. A dead organism or any product derived from it can no longer receive radioactive carbon from the atmosphere and therefore the radioactivity of its carbon must diminish at a steady rate. Therefore any prehistoric artifact whose carbon has one half as much radioactivity as a contemporary sample must be 5,900 years old. (The contemporary levels are those determined prior to the detonation of atom bombs. Since 1945 the C^{14} content of the atmosphere and of living things has been increasing.)

If we wish to determine the age of the Dead Sea Scrolls or of the cave paintings at Lascaux we can do it with ease by burning to carbon dioxide a tiny sample of the cloth wrappings on the scroll or the charcoal which was used as paint by the prehistoric frescoist. The carbon dioxide is fed into a Geiger counter, whose clicks unerringly count off the centuries elapsed since the cloth or the charcoal was part of a living thing.

Isotopes are a wonderful tool of detection. With their aid we can follow the path in the body of a particular hydrogen atom in a particular molecule. Questions which seemed unanswerable in pre-isotope days are being answered routinely. Most of our knowledge of the intermediate metabolism—the over-all fate from absorption to excretion—of fats and amino acids we owe to isotopes.

But isotopes won't yield miracles. When the newspapers and magazines became isotope conscious they began to predict medical millennia just around the corner, produced by isotopes. The dragons of cancer, heart disease, or whatever ails one were being slain by the isotopic swords. While making the public research conscious is extremely valuable, it is cruel to raise false hopes. Isotopes are a tool, a good tool, but just one of many tools.

A tool by itself has never built anything. The scientists whose minds and hands wield the tools are the architects of medical research. Only the ideas of men and women who can dream them will penetrate the startling complexity of a living cell. Only when we have a clear view of the normal pathways in the labyrinth of the cell will we be able to trace the monstrous blunders which lead into the cul-de-sacs of cancer or arteriosclerosis.

Bold, direct frontal assaults have been made on cancer, using radioactive isotopes as weapons. But we cannot as yet control the range of the weapons; it is shooting friend and foe alike. The sensitivity of cancer cells to high-energy radiation such as X rays and radioactivity is well known. However, normal cells are vulnerable too, and unfortunately the staying power of the normal cell under the impact is only very slightly greater than that of the wildly growing cancer cells. There is always some destruction of normal tissues as well.

The dream is to send packages of radioactivity by special delivery into only the diseased cells. We are back to Ehrlich's old problem: to carry the "magic bullet" to specific cells. This time the bullet has, not arsenic, but radioactivity in its warhead; otherwise the problem has not changed. As the French say, "The more things change, the more they remain the same." We can but hope that there is another Ehrlich, not too far away, who will direct the bullet to its mark before the biological scientists unravel the labyrinth of the cell.

6. AMINO ACIDS AND PROTEINS

MASONRY OF OUR CELLS

"LET THEM EAT CAKE," was the reputed dietary advice of Marie Antoinette to the undernourished poor of Paris.

Came the revolution, the poor, instead of more food, got more advice: "Let them eat glue." This dietary exhortation came from Cadet de Vaux, a physician, who urged Parisians to make soup out of glue, or gelatin, guaranteeing it to be as nutritious as beef soup. The government issued official proclamations endorsing the new substitute for Marie's cake. The Institute of France and the French Academy of Medicine added their authority, praising the ersatz food and cajoling the starving Parisians to become converts to it.

But the Parisians would have nothing to do with such newfangled nonsense. A political revolution they took in their stride, but a revolution of the stomach, that is a serious matter.

About a century and a half later biochemists proved the wisdom of the adamant Parisians. There were attempts to evaluate the food value of gelatin long before the biochemist brought some order to the chaos of the field of nutrition. The most interesting attempt in this prehistoric era of nutritional science was made by M. Gannal, a manufacturer of glue who boldly resolved in 1832 to test the food value of his product. He noticed that the rats which infested his factory ate the raw materials—the tendons, cartilage, and skin of animals—but snubbed his product, the glue, which was obtained by cooking these animal wastes with water.

Gannal conjectured that perhaps rats were merely fastidious about the taste and odor of glue. He therefore decided to perform his feeding experiments on humans. He is to be commended for not urging the consumption of his product on the poor; he fed it to his wife and three children and he himself joined them in their dreary diet. They ate glue, and, to make it somewhat more palatable, glue and bread, for weeks. The result was disastrous. They had violent headaches and intense nausea, and when it became apparent that their health was rapidly deteriorating, M. Gannal reluctantly called off the experiment. He sadly concluded that his product has no food value, indeed it is harmful.

The only thing more difficult than the introduction of a new, fruitful idea, is the banishing of an old, fruitless one. The feeding of glue, or in its more purified form gelatin, kept cropping up in scientific and medical literature for the next hundred years. Convalescents in hospitals, nursing mothers, and infants were fed gelatin. Fortunately, in most cases the feeding period was brief. Nor was gelatin the sole protein in the diet. The conclusions drawn from these improvised experiments were varied, depending upon the susceptibility of the subjects and of the experimenters to autosuggestion.

The field remained chaotic until biochemists entered it with their zoos of experimental animals. First of all, it was established that proteins are an absolute essential in the diet of rats and of dogs. Caged animals, which could not forage for food, were kept on diets completely devoid of proteins. They rapidly lost weight and, unless they were rescued with meals of proteins, they died.

The next question was: Is it the proteins themselves or their amino acids that are essential to the animals?

A wholesome protein, casein from milk, was cooked with acid until it was whittled down to its amino acids. Rats which had no protein in their diet were fed this mixture of amino acids. The rats thrived. This should not be surprising; after all, what happens to the proteins in the alimentary canal of animals? They are broken down by the enzymes into amino acids. Animals do not absorb whole proteins into their blood streams; they absorb only the amino acids. Whether the protein is crumbled down by acid in the flasks of the chemist or by enzymes in the stomach and intestine of the animal apparently makes no difference.

Progress in science is similar to a duel with the mythical multiheaded Hydra. For every question answered, other new questions crop up. Once the biochemist proved that it is the amino acids, not the whole protein, which is essential, he had to tilt with a brand new and far more difficult question. Are all amino acids of equal value to the animal? The obvious way to resolve this question is to feed all of the amino acids to experimental animals and then keep withdrawing the amino acids from such a diet, one at a time.

While this scheme of eliminating amino acids is simple and obvious it could not be carried out until recently because of technical difficulties. Individual amino acids can only with difficulty be cajoled out of a natural mixture of them. They are so alike in chemical behavior that they manage to hide each other from the chemist who is bent on extracting one of them.

Biochemists therefore turned to some natural proteins which were known to be lacking one or more amino acids. Gelatin was found to be such an incomplete protein. Three amino acids are completely missing from it and two others barely put in an appearance. Gelatin is by no means the only protein so poorly endowed with amino acids. Zein, a protein from corn, and gliadin, a protein from wheat, are both lacking in some amino acids.

Rats cannot live on a diet complete in every other way but containing as the only source of amino acids these impoverished proteins. Their growth is stunted, and unless help comes in time, they die. The help is either a wholesome protein like casein or the original, incomplete protein fortified with the missing amino acids.

The growth curves of rats on such a diet illustrate vividly the need for these amino acids. Weaned, litter-mate rats are used in such experiments to rule out individual variation.

A rat on a complete diet containing casein as the protein grows like this:

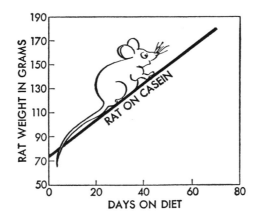

If in the same nutritious rat meal casein is replaced by zein, the protein from corn, the rat will lose weight at an alarming rate.

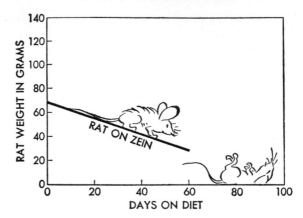

But if we come to the rescue of the moribund rat and throw him a life belt of the two missing amino acids, tryptophan and lysine, the rat will perk up and will begin to gain weight with sufficient speed to walk right off the graph.

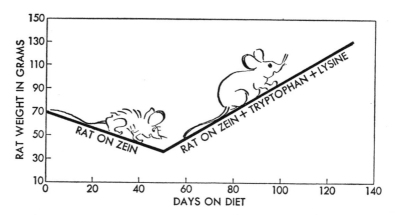

The next question in this logical sequence was: Does the rat find every single amino acid essential in its diet? Very little work was required to answer this question. Casein is one of the best proteins; all suckling animals thrive on it. Yet this wholesome protein lacks some amino acids. We must conclude, therefore, that all amino acids are not essential, only some of them are.

But which of the twenty amino acids are essential? Had there been twenty different proteins each lacking a different acid, our task would have been easy. Those proteins which could not support growth, we could conclude, lack an essential amino acid. Unfortunately, the various proteins found in nature were not designed for the convenience of the biochemist. There are no such proteins with prefabricated amino acid deficiencies. We had to sort out the essential and nonessential amino acids the hard way.

Biochemists have been nibbling at this problem from the start of the century without making too much headway. About thirty years ago William C. Rose, a biochemist at the University of Illinois, started a series of well-designed experiments which were to serve as a powerful beacon illuminating this whole complex field. We learned, under the light of this beacon, that there are no less than ten essential amino acids and that a protein is worth no more than the amount of these amino acids it contains. Dr. Rose decided to keep rats on a collection of pure amino acids rather than on proteins. In this way he could leave out one amino acid at a time and study the effect of the omission on his rats.

It is difficult to convey to the reader what an onerous task this was. Growth studies are long drawn out affairs. In some cases rats had to be continued on the same diet without a day's interruption for six to seven months. To test twenty different amino acids meant keeping twenty different groups of rats on twenty different diets.

The pedigreed rats [1] used by the biochemist are sensitive creatures; they react to the slightest changes in their environ-

---

[1] The genealogy of these rats is better known than that of the noblest entrant in the Almanach de Gotha.

ment. Noise, variations in temperature, changes in lighting, the freshness of their food, all affect their growth rate.

Only a few of the amino acids were available commercially, and those were prohibitive in cost. Since these experiments were conducted years before the atom bomb became the exclamation point to the scientist's plea that research pays, money for research was not easily forthcoming.

Rose and his graduate students at Illinois labored long at the accumulation of amino acids. Some they isolated from mixtures of amino acids obtained from proteins; others they made synthetically in the laboratory.

When all the known amino acids were assembled they were fed to rats which were receiving no proteins at all. The rats failed to grow. They could be induced to grow only if the amino acid mixture was supplemented with a bit of casein. Dr. Rose concluded that something else besides the twenty known amino acids must be present in casein and that this unknown substance is also essential for the rats. To track down the unknown factor he began one of those tedious, painstaking searches which the reader by now must recognize as an occupational hazard of the biochemist.

Casein was cooked with acid and the resulting amino acid mixture was put through a variety of chemical separations. Each of these fractions was fed to rats along with their diet of amino acids. One fraction when added to the diet of the known amino acids enabled the rats to grow. This fraction, after appropriate purifications, yielded a brand new amino acid, threonine, of whose existence we had not even dreamed, so well had it been hiding among its brethren. Rose started once again to feed his rats all the previously known amino acids plus threonine. The rats thrived.

Then he started a group of rats on a diet containing all the amino acids but one, and he measured their growth. Another group of rats was put on a diet lacking another one of the amino acids and this was repeated until all of the twenty-one amino acids had their turn at being left out in this most elaborate of musical chairs games.

Ten different groups of rats failed to grow; ten amino acids are essential to the rat. They can do without the other eleven amino acids, but if just one of these ten essential amino acids is missing the rat behaves as if it were receiving no protein at all; it cannot grow.

The need for protein, then, is the need for these ten essential amino acids. "By their essential amino acids shall ye value them," should be the injunction to guide us in our choice of proteins. For it has been found more recently, that in the need for the essential amino acids, we humans are akin to the rat. We are not quite as exacting as the rat in our dietary requirement. Our cells can make one of the amino acids and thus we require only nine of them as absolute essentials in our diet.

Now we know why the Gannal family became ill on their diet of gelatin: they lacked some essential amino acids. However, no harm results from eating gelatin. The writer would not recommend gelatin as the sole source of protein in the diet, but as a low-caloried, decorative adjunct to a meal it is useful.

The distribution of the essential amino acids in foods is unfortunate. We find that only the expensive animal proteins —meat, cheese, and eggs—are well endowed with them. The cheaper plant proteins are either low or completely lacking in some of the essential amino acids. Some bean proteins, especially those of the soya bean, approach, but never reach, the animal proteins in their essential amino-acid value.

The impoverished nature of vegetable proteins is interesting in view of the fairly widespread fad for vegetarianism. In the recent presidential elections the Vegetarian Party ran candidates for that office. The enticing program this party offers is the slaughter of all food animals, and the conversion of all grazing lands into farms to raise vegetables in sufficient abundance to make up for the loss of meat. Whoever concocted this little scheme knew nothing about nutrition and even less about agriculture. The production of meat is by far the most efficient, indeed the only way to utilize over a billion acres of our marginal land. Close to a billion acres of ranch land are not fit for raising anything but grass. This land would be totally unproductive for human nutrition without grazing animals, which alone can convert the grass into meat. Of another half a billion acres of pasture land about three fourths are too hilly for plowing. (We must concede one virtue to the vegetarian platform: it is explicit, a rare quality in this very specialized kind of literary endeavor.)

The practice of absolute vegetarianism from early childhood would be disastrous. The only reason that the effects of the poverty of essential amino acids do not become apparent in vegetarians is simply that there are no absolute vegetarians. The possible exception is the Chinese coolie with his daily bowl of rice, and he is not noted for robust health. Most vegetarians eat milk, cheese, and eggs.

Even the most celebrated vegetarian, the late George Bernard Shaw, had strayed from the true path: he used to take liver pills. He referred to these furtive adjuncts of his vegetarian diet as "those chemical pills."

Partisans of vegetarianism had triumphantly ascribed Mr. Shaw's zestful long life to his partially vegetarian habits. Advocates of health fads indulge all too frequently in this kind of illogic. Well-controlled experiments on the human diet, running for decades on the same individuals, are impossible to achieve. But hope for longevity is fervent, and it is so easy to draw dietary conclusions to nurture that hope.

Those bent on partisan explanations of longevity are particularly apt to commit the "after this, therefore because of this," error of logic. Everything a man has ever done preceded his old age.

The amounts of essential amino acids needed are fortunately so low that a person on an average, well-balanced diet can dismiss all concern about them. For example, the British Ministry of Health advocates the daily consumption of not quite two ounces of a "first-class protein," meaning meat, fish, cheese, or eggs. American authorities advocate about two and a half ounces of mixed proteins per day for an adult. Obviously anyone who eats an egg or two and a fair serving of meat or fish a day is well provided with "first-class protein." Hypoproteinosis, the condition induced by lack of good proteins, is seen only among people on very impoverished diets. Its symptoms are somewhat indistinct; vitamin deficiencies and the cumulative effects of squalor are invariably associated with it.

A multitude of functions in our bodies make the amino acids so indispensable to us. In the first place, they are the bricks of which our tissues are built. Twelve of these twenty-one bricks can be made in our own bodies; that is why they can be left out of the diet with impunity. If one of these twelve amino acids, for example, alanine, is omitted from the diet, we can make enough of it for our needs.

One of the stages in the metabolism of sugars, is pyruvic acid. Alanine, the amino acid for whose absence we are about to compensate, and pyruvic acid are very similar in chemical structure. They both consist of a chain of three carbon atoms, two of which—the two end ones—have the same atoms attached to them. They differ only in the atoms that the middle carbons bear. In pyruvic acid there is an oxygen atom attached to the middle carbon, in alanine an atom of nitrogen.

If we happen to need some alanine to build into our body proteins and none of it is forthcoming from the diet, the enzymes in our liver improvise the alanine. They marshal a molecule of ammonia and a molecule of pyruvic acid and clip them together to form alanine. The oxygen atom, which is set free from the pyruvic acid, soon gathers up two hydrogens and they swim away as a water molecule.

Thus we can make an amino acid out of a by-product of sugar metabolism. This is a great asset. We are not slavishly dependent upon our diet for twelve of the amino acids. We can produce each of these by making over some fragment from the utilization of our sugars and fats, upholstering such a fragment with ammonia. But the nine essential amino acids that we cannot make must come to us ready made. We can do a bit of an assembly job, we can add ammonia to the appropriate carbon skeleton, but we cannot fabricate the carbon skeleton of the nine essential amino acids. As the mason needs a specially designed keystone when building an arch, so our enzymes need the nine prefabricated essential amino acids for building proteins.

Nor is the assembly into protein molecules the only role of amino acids. They are the most versatile molecules. They are made into hormones and into body pigments, and they are unleashed to disarm poisonous invading substances.

It is beyond the scope of this book to describe the special role each amino acid performs in the body, but one, methionine, will be used as an example. Methionine was chosen for two reasons: it has an interesting history and it performs interesting functions in our cells.

Methionine makes up 3 percent of casein and it is essential

to rats, men, and microorganisms. Yet we had no idea of its existence until 1922, or of its structure until 1928. It is another one of those biologically important substances which was discovered by work not on animals but on microorganisms. It was discovered by the late Dr. John Howard Mueller, whose chief interest was the dietary requirement of diphtheria bacilli.

This aspect of bacteriology, the nutritional need of microorganisms, was in chaotic disarray at the time. Every science passes through such a muddled period, until coordinating principles are found to weave some pattern out of the mass of apparently unrelated bits of information gathered by the pioneers of the science. Chemistry did not emerge from alchemy until the end of the eighteenth century; bacteriology, a much younger science, started to emerge from its chaotic morass in the second decade of this century. Bacteriologists had tried everything at frantic random to grow bacteria away from living host animals. The prescriptions for raising bacteria, until recently, read like a cookbook for apprentice witches. A concoction might be made of a pig-heart infusion, with a dash of yeast extract, and a soupçon of beef blood. Today, many microorganisms can be grown on completely synthetic diets. As for the others, just give us time, they will eat out of our hands yet—synthetic food.

Mueller was interested in the amino acids needed by diphtheria bacilli for growth. Instead of the bacteriological witches' brew he raised them on acid-cooked casein. He then put the amino-acid mixture through the usual chemical fractionations and tested each fraction as a diet for the bacilli.[2]

From one of these fractions he isolated a hitherto unknown amino acid, methionine, which proved to be essential for the diphtheria bacilli. About ten years later Rose found the same

[2] This is of course the very process Rose was to repeat about ten years later in his search for the amino acids essential to the rat.

amino acid essential for the rat and only a few years after that the amino acid was being used in the therapy of some diseases of the human liver. The history of methionine offers but one more example of the unforeseen bounties that can accrue from basic research. That not only the public, but even pharmaceutical houses were blind to the value of research is all too clear from Mueller's experience. Recently he wrote: "The writer recalls somewhat grimly the difficulties encountered in 1920 while attempting to enlist cooperation [of pharmaceutical companies] in getting a hundred pounds of casein hydrolyzed with sulfuric acid, from which methionine was eventually isolated." The current profit from a day's sale of methionine would more than cover the cost of the little favor for which Mueller was pleading.

What does methionine do that makes it indispensable to man, beasts, and bacilli? Methionine is one of the two commonly occurring [3] amino acids which contain, in addition to the usual elements, the element sulfur. The fate of the sulfur of methionine in animals has been charted with the aid of isotopes. Methionine has been made in which the sulfur atom is not an ordinary sulfur but a radioactive isotope. This sulfur can be traced by the radioactive messages it sends to a receptive Geiger counter. It turns up in the other sulfur-containing amino acid, cystine, which makes up almost 15 percent of the proteins of the skin and hair. These are very special proteins: they are tough and they are insoluble in water. We should be especially grateful to some creeping ancestor of ours who first

---

[3] There are more than twenty-one naturally occurring amino acids. Quite a few others are found in more rare sources; for example, the octopus yields octopine. While these rare amino acids apparently play no role in mammalian nutrition, we treat them with respect. One such amino acid was found by a Japanese investigator in watermelon seeds. A few years later this amino acid was found to be not so rare after all: it plays a stellar role in the formation of urea in the human liver.

acquired such a skin by a fortuitous mutation. Imagine having a protein such as egg white for the skin. Getting caught in a rain would be fatal; our tissues would trickle away and nothing would be left but our bones, in the midst of a puddle of tissues.

One of the roles of methionine, then, is to provide the raw material sulfur, for one of the amino acids of the skin (and of other tissues as well).

Attached to the sulfur atom in methionine is a very simple group of atoms—one carbon, loaded with three hydrogen atoms —called the methyl group. It is the simplest pattern in organic chemistry. The methyl group of methionine, too, has been traced with the aid of isotopes. After methionine is eaten, the labeled methyl groups from it turn up in several different substances in the body, substances of vital importance: for example in the hormone adrenalin and in choline, a compound with many roles. Choline aids in the metabolism of fats; it is also part of the impulse-sending mechanism in nerve tissue. It contains three methyl groups in its architecture. These methyls are supplied by methionine.

A deficiency of methionine and of choline can be disastrous for the animal. Its liver becomes diseased and degenerated. Choline and methionine have become valued tools in the treatment of certain diseases in the past few years. From the curiosity of a bacteriologist about the dietary needs of the diphtheria bacilli, a medication for humans was harvested.

And what happens to the remnants of the methionine molecule? After detaching the methyl group and the sulfur atom, we are left with four carbons to which is attached a nitrogen atom. All we know about this fragment is that its nitrogen can be severed from the rest of the molecule to form ammonia which, in turn, can be incorporated into urea and then excreted. What happens to the rest of the molecule is still unknown. It can be converted into glucose and thus serve as a source of energy; but whether this part of the methionine molecule has any unique role we do not know.

Finally, methionine as a whole enters into a combination with other amino acids to form that most marvelous of substances: the protein of our tissues. The protein molecule is one of nature's masterpieces of complexity. In the elaborate pattern of that molecule is locked the secret of life. The proteins are the mechanics of life: they fabricate its tissues, regulate its energy, and assure its perpetuation. How the amino acids are lined up faultlessly to form these molecules with almost magic attributes will be the subject of the next chapter.

There is no extensive storage of proteins in our bodies. We can store huge amounts of fats in our fatty depots and we store sugar in the glycogen depots, but we store no proteins in excess of the amounts needed to make up our tissues. On the one hand this is a serious drawback, since we must, therefore, depend on a steady supply of essential amino acids from our diet. But then, many proteins are active enzymes; indeed some of us think that every protein molecule is an enzyme. The unchecked accumulation of highly potent protein molecules could become disastrous: we might grow to monstrous size and an uneven distribution of proteins in our bodies could wreak havoc with the delicately adjusted balance of the various parts of the whole individual. The accumulation of large blobs of fat around our abdomen does damage only to our vanity. The deposition of similar amounts of protein would be lethal. Even hibernating animals, which must lay in a large store of food in their bodies for a long siege of starvation, store only fats, not proteins.

If an animal happens to receive an overgenerous supply of amino acids, in excess of its immediate needs, it metabolizes them into urea, carbon dioxide, and water; or, if it requires no immediate source of energy, it can convert the excess amino acids into sugars and fats, the depots of which can be readily augmented. Earlier in this chapter the process by which an intermediate in sugar metabolism, pyruvic acid, can be converted into the amino acid alanine, was outlined. If, on the other hand, an animal is encumbered with an excess of alanine, it reverses this process: it converts the alanine into pyruvic acid. Certain enzymes—mostly in the liver and kidney—remove the ammonia from the amino acid and replace it with an atom of oxygen. Other enzymes combine the newly formed ammonia with carbon dioxide to form urea which is then excreted. Thus two fairly toxic waste products, ammonia and carbon dioxide, are efficiently eliminated in one step. The molecule of pyruvic acid is either further metabolized to carbon dioxide and water or two molecules of it are fused together to form a molecule of glucose. The pyruvic acid can, by a more elaborate process, be converted into a fat, too. Thus the proteins in our diet are ready sources of carbohydrate and fat. It is well to remember this, since the nonfattening nature of proteins has been overemphasized by the designers of reducing diets. While it is true that proteins have lower caloric value per unit weight than fats or carbohydrates, nevertheless, unfortunately for those bent on acquiring a more stylish figure, a thousand calories coming from lamb chops are just as fattening as a thousand calories from rice pudding.

Architecture is frozen music.
FRIEDRICH VON SCHELLING

7. ATOMIC ARCHITECTURE

THE STRUCTURE OF INSULIN

THE ROCKET that hurled Russia's first Earth satellite into orbit also served as a starting gun for a marathon debate on American education. Hardly a day passes that we do not read in our newspapers of demands for more sciences in our curricula, or of counterpleas: "Let us not neglect the humanities."

Josiah Willard Gibbs, the great American theoretical scientist of the nineteenth century, once attended a faculty meeting at Yale where a debate droned on about the relative merits of mathematics versus languages in an undergraduate curriculum. Apparently then, as now, nothing aroused members of a faculty to greater heights—and lengths—of eloquence than the carving up of a student's academic carcass. Gibbs, who was retiring to the point of being a recluse, listened to the harangues about the superior merits of languages. Finally he rose and is reputed to have said: "Gentleman, mathematics *is* a language" —and left.

One is tempted after hearing much of the current debate on the humanities versus the sciences to paraphrase Willard Gibbs: "Gentlemen, the sciences *are* humanities." They are but slightly different branches on which creative imagination blooms. Their seed is nurtured by the spirit of the time; their fruits are savored by generations; their impact is sometimes identical.

In 1857 two Frenchmen, a novelist and a chemist, published their major works. Flaubert with his *Madame Bovary* launched the modern novel of realism; Pasteur with his study of fermentation started us on a path of realism in dealing with life's processes. The one cast away the idealized images of human emotions and behavior, the other lifted the miasma of vitalism which befogged our view of biological processes.

Even aesthetically a creation of science can approach man's noblest works of art. It is only unfortunate that the aesthetics of science has a very limited audience. Any one with sight and soul can enjoy a Bernini statue, but only those versed in the language of science can savor the beauty of a scientific creation. Some years ago I overheard an instructor at Cambridge University extolling to a group of undergraduates the beauties of King's Chapel. The teacher felt this was the ultimate in architectural beauty, never to be surpassed or even equaled by another work of man. Neither the instructor nor his students were aware that at that very moment, only a stone's throw away in a poorly equipped laboratory, a young biochemist was performing an architectural feat of such subtlety and beauty that his work, I felt, was more than a match for King's Chapel. The creator of that noble vaulted structure and the biochemist wielded different tools, worked in vastly different dimensions, but each brought to his task the ultimate in the skills and the knowledge of his century and, as genius will, each leaped ahead of his contemporaries to create an enduring triumph of his time, the one in stone, the other in atoms.

To one who understood him, listening to Dr. Frederick Sanger unfold his tale on the decoding of the structure of insulin was just as thrilling as King's Chapel at twilight with the organist weaving a Bach fugue. I shall try to convey to the reader the extraordinary ingenuity with which the problem of the struc-

ture of the protein hormone insulin was attacked and the sheer
beauty with which it unfolded.

Proteins are composed of long chains of amino acids. The
links of amino acids are forged together by the electronic forces
of the atoms. To achieve a union, two amino acids together
shed a molecule of water. Part of the water molecule comes
from the acidic group of one amino acid, part from the amino
group of the other. The electronic forces which originally had
held onto the atoms which form the water now graft the two
amino acids together. In other words, an atom fragment is
carved out of each amino acid and fusion takes place at the
shorn sites. Such a couplet of amino acids still has the amino
group of the first amino acid and the acidic group of the second
unengaged. These can reach out to two more amino acids and
so they continue until vast accretions accumulate. Some proteins are as much as a million times as heavy as a hydrogen
atom. The individuality of each protein chain is shaped by its
sequence of amino acids. From that sequence are derived the
physical and biological attributes characteristic of each protein.

Since there are twenty different amino acids, the variety of
protein molecules which can be woven out of them is beyond
imagining. Let us look at a simple example.

If only the three amino acids A, B, and C are used to form
a triplet—a so-called tripeptide—the following are the different
products possible:

A B C
A C B
B A C
B C A
C A B
C B A

The permutations possible can be predicted without listing them. It is factorial 3, symbolized as 3! which is equal to $1 \times 2 \times 3 = 6$.

In longer chains of amino acids the number of possible variations fairly leaps into astronomical magnitude. With twenty different amino acids the permutations possible are 20! which is equal to two billion billions (2×10^{18}). And such a chain would be puny as proteins go, weighing only about 2,000 times as much as one hydrogen atom.

The way in which amino acids are linked together has been known for over fifty years from the work of the great German chemist Emil Fischer. But until fifteen years ago it appeared that the decoding of the *sequence* of amino acids in a protein would present unsurmountable difficulties. That such a feat has been accomplished is a tribute to the ingenuity of three English chemists, Dr. A. J. P. Martin, Dr. R. L. M. Synge, and Dr. Frederick Sanger.

Doctors Martin and Synge provided an extraordinarily simple analytical tool for the detection of the presence of each amino acid. Prior to their achievement, the analysis of amino acids was a grueling task with uncertain outcome. Amino acids are so similar in chemical and physical behavior that their separation in pure form—a *sine qua non* of analysis—was tedious and incomplete. For example, twenty years ago a well-known chemist completed the first reliable assay of glutamic acid in the protein of milk. It took him a year to complete his study, and he had to start with a hundred grams of the protein. Today a skilled assistant can complete a dozen assays of glutamic acid in twenty-four hours by paper chromatography, and he needs but one millionth as much of the starting material. This technique, which did for protein chemistry what the plow did for farming, is based on a very simple principle. It depends on the fact that liquids will rise, defying gravity, along thin fibers or tubes. If a large sheet of filter paper is rolled up and the lower edge of it is inserted into a dish of water the moisture will creep up the paper. If kept in that position for twenty-four hours, the water will rise to a height of from twelve to fifteen inches. Martin and Synge noted that if minute amounts of chemicals, say amino acids, are placed on the bottom of such paper they will migrate upward with the liquid; the relative height to which they travel is an unvarying characteristic of each amino acid. There is a simple reagent which when sprayed on the dried paper will reveal the area to which the amino acid had risen by turning the spot lavender. We thus have a delightfully easy method for the detection and identification of as little as a hundred thousandth of a gram of an amino acid. We can compare the rate of migration of an unknown substance with that of a genuine sample of an amino acid, and if they travel together and give the same color reaction, they are identical.

Doctors Martin and Synge received the Nobel Prize for their simple scheme, for they opened up frontiers of biochemistry which had previously remained barred to clumsier analytical tools. For example, we can extract from fossils minute traces of organic matter and ask them via paper chromatography what they contain. The astonishing answer from these remnants of creatures that lived as long ago as three hundred million years is that their bodies were constructed of the same amino acids as ours.

With this tool in hand Dr. Frederick Sanger boldly ventured on a vast enterprise: to determine the structure, that is, the exact sequence of amino acids, of a protein. He chose insulin for the first assault for three reasons: (1) it is the smallest protein known with definite physiological properties; (2) it was available in pure crystalline form; (3) it was relatively cheap, an important consideration for a chemist working on a minuscule budget. (In Dr. Sanger's laboratory washbowls long discarded from Victorian bedrooms served as makeshift scientific utensils.)

It was known from the probings of several chemists that insulin contains only sixteen of the known amino acids. It was also known that some of these amino acids are repeated as many as six and seven times. The average weight of an amino acid is 100 times that of hydrogen, therefore a chain of sixteen amino acids would weigh only 1,600. However, insulin weighs 6,000. Therefore, some of the amino acids must be repeaters in the structure.

With this information at hand Dr. Sanger was ready to tackle the most intricate cryptogram ever to challenge the human mind. His first goal was to devise a method of pinpointing the first amino acid in the chain. That is the one whose amino group is not forged into a link. There had been previous attempts to achieve such a labeling, but none were too successful. An agent for such a task has to be a molecule which seeks out only a free amino group and which adheres to it so tenaciously that the amino acid chain can be dismembered by heating with acid without removing the label. If in the debris of amino acids the one with the label can be found, we can be certain we have the head of the chain. Dr. Sanger devised a remarkably effective label which meets all the requirements and has an added fillip: it is bright yellow. It thus serves as a flag amidst the amino acids, all of which are white. Once this compound, now called Sanger's reagent, was on hand decoding could start.

Insulin and Sanger's reagent were permitted to unite, and the yellow product was dismembered with strong acid. The solution of amino acids was then subjected to paper chromatogra-

phy. The sixteen amino acids arrayed themselves as antici-
pated, but on inspection they revealed a surprise. There were
not one but two different amino acids carrying Dr. Sanger's
label. Therefore there must be two amino acids—they turned
out to be glycine and phenylalanine—whose amino groups are
not engaged in the weaving of the insulin molecule.

This intelligence could be interpreted only one way: insulin
is not a single strand of amino acids; it must be a double one.
How are these two strands held together? Insulin is rich in the
sulfur-containing amino acid cysteine, which has a unique
attribute. It has not two but three hooks with which it can
form protein links. In addition to its amino and acidic groups
its sulfur atom can reach out to the sulfur atom of another
cysteine in a neighboring strand and forge a fairly strong link.

Our knowledge up to this point can be summarized in the
accompanying diagram, in which a flag designates Sanger's
reagent (the exact position of the cysteines is unknown).

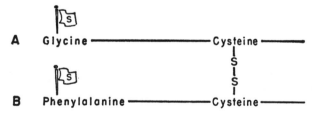

The next order of business was to separate the two chains
(A and B). This was relatively simple, since the sulfur links
were known to be easily broken by oxidation. A mixture of the
two chains was thus obtained, but they had to be separated. At
this point Dr. Sanger's ingenuity was matched by good luck:
the two chains turned out to have different electrical charges
on them. Molecules with different charges migrate in different
directions in an appropriate electrical field, and we are there-

fore able to separate such a mixture into two homogeneous components.

The two different chains produced by the cleavage of the sulfur bridges were separated and purified. Chain A, which had glycine as its first amino acid, had about twenty amino acids in it; chain B, whose guide bearer was phenylalanine, had about thirty.

Chain B was tackled first because it promised to be easier. There are a larger variety of amino acids in it, therefore there was a likelihood that some of them might appear only once. In chain A, with a smaller variety of amino acids, the probability of repeated appearance is greater, which places an extra burden on the decoding.

Chain B was treated with Sanger's reagent and was subjected to very mild fragmentation. This is done with weak acid and very gentle heating. Under these conditions the cleavage is incomplete, yielding fragments of the chain containing two, three, and four amino acids still hooked together.

Such a mixture of debris was paper chromatographed and allowed to migrate according to the propensities of the components. (Their rate of travel is slower than that of individual amino acids.) Dr. Sanger observed several different yellow spots. Any labeled phenylalanine which was free of all other amino acids formed one spot. A phenylalanine which still had the next amino acid attached to it, formed another spot, one with two amino acids attached to it still another, and so on.

Each of these spots on the paper was cut out, the amino acid chains leached out of them, and each solution evaporated. Dr. Sanger now had a series of fragments of the original chain.

Each of the leached-out spots was cooked with acid to break them down completely, and they were analyzed by paper chromatography for the identity of the amino acids.

Dr. Sanger was successful in finding fragments which contained only two amino acids, the labeled phenylalanine plus another amino acid. It turned out to be valine. Now the decoding was under way. The first two amino acids of chain B are: phenylalanine-valine. A triplet of amino acids was also found which contained labeled phenylalanine, valine, and a third amino acid, aspartic acid. Therefore, the sequence of the first three amino acids must be

Phenylalanine—Valine—Aspartic Acid

1   2   3

A quadruplet of amino acids revealed the sequence:

Phenylalanine—Valine—Aspartic Acid—Glutamic Acid

1   2   3   4

A quintuplet of amino acids yielded the following sequence:

Phenylalanine—Valine—Aspartic Acid—Glutamic Acid—Histidine

1   2   3   4   5

This was the longest of the chains still containing labeled phenylalanine which Dr. Sanger could find. At this point he resorted to other tactics. He turned to the amino acid complexes which did not have the labeled number one attached to them. In other words, he started examining the mixtures which originated from further along the chain. He found a triplet which contained amino acids 4 and 5 and a new one which was not 3. A chain of thirty amino acids can form only ten triplets. Since there are sixteen different amino acids to be distributed among those ten triplets, the probability of 4 and 5 appearing more than once is very small. Therefore the triplet containing 4 and 5 and the new one most likely was 4—5—6. Amino acid 7 was found by obtaining a triplet containing 5—6 and a new one.

Dr. Sanger continued hunting and matching overlapping fragments with the sensitivity of an artist, the concentrated intensity of a genius, and the ingenuity of a truly original mind.

The sequence of some of the fragments from further along the chain could be decoded only by labeling *their* first amino acids with Sanger's reagent and performing a stepwise fragmentation on them. As the sequence was laboriously wrested from each fragment it was fitted into the larger pattern until, finally, thirty amino acids were unequivocally delineated.

He then tackled chain A and, using the same strategy, decoded a sequence of twenty-one amino acids.

He next attacked the question of the location of the sulfur bridges through which the two chains are connected. It is easier to visualize his approach if we line up the two chains side by side, designating the amino acids by numbers (see Figure 1).

In chain A cysteine appears four times: 36—37, 41, 50. In chain B it makes its appearance only two times: 7, 19.

The question is which of the six sulfur atoms are the abutments through which the bridge or bridges are formed. To answer this question Dr. Sanger's ingenuity once again provided a unique approach. He subjected whole insulin, with the sulfur bridges intact, to disintegration by enzymes which were known not to sever sulfur bridges, and he started once again to hunt for fragments. This time he sought only fragments which contained cysteine. He found novel sequences which exist neither in chain A nor B. For example, he found a fragment of four amino acids, as in this diagram:

$$
\underset{\text{Glycine}}{8} \underset{\text{Cysteine}}{7} \underset{\text{Cysteine}}{37} \underset{\text{Alanine}}{38}
$$

Since such a sequence appears in neither chain A nor B, the two cysteines must come from the two different chains carrying a neighbor with them; therefore, there must be a bridge between 37 of A and 7 of B.

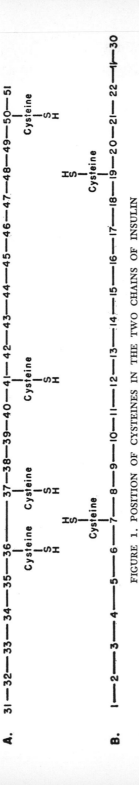

A.

31 —32—33 —34 —35 —36 —37—38—39—40—41—42—43—44—45—46—47—48—49—50—51

Cysteine   Cysteine Cysteine   Cysteine   Cysteine
| | | | |
S   S S   S   S
| | | | |
H   H H   H   H

B.

1 —2 —3 —4 —5 —6 —7 —8 —9 —10—11—12—13—14—15—16—17—18—19—20—21—22— 30

Cysteine   Cysteine
| |
H   H
| |
S   S

FIGURE 1. POSITION OF CYSTEINES IN THE TWO CHAINS OF INSULIN

A.

NH₂   S   NH₂   NH₂   NH₂

Gly · Ileu · Val · Glu · Glu · Cy · Cy · Ala · Ser · Val · Cy · Ser · Leu · Tyr · Glu · Leu · Glu · Asp · Tyr · Cy · Asp

S   S

B.

NH₂ NH₂

Phe · Val · Asp · Glu · His · Leu · Cy · Gly · Ser · His · Leu · Val · Glu · Ala · Leu · Tyr · Leu · Val · Cy · Gly · Glu · Arg · Gly · Phe · Phe · Tyr · Thr · Pro · Lys · Ala

FIGURE 2. THE STRUCTURE OF INSULIN

Similarly, he found another bridge between 50 of A and 19 of B.

Thus, after a decade of intense dedication Dr. Sanger could unfurl one of the proudest pennants signaling a human achievement, the structure of a physiologically active protein (see Figure 2).

The structure of insulin reveals no periodicity of any kind of its amino acids, but rather a random sequence. Somewhere in that sequence the physiological potency of this hormone must reside. Where, exactly, we do not know. Part of the chain, at least, can be altered without impairing the function of the hormone. All of Dr. Sanger's original work was done on beef insulin. When, more recently, the insulin of four other species of mammals—the pig, the horse, the sheep, and the whale— were examined, their structures contained a surprise for us. Chain B was the same in each. But in chain A the amino acids in sequence 38—39—40 were different in each species. The amino acids in that part of the chain seem to serve merely as struts to hold the structure to a definite dimension and shape. Interestingly, the structure between amino acids 36 through 41 appears in two other hormones which are also composed of amino acids, oxytocin and vasopressin. What, if any, significance this has we have no idea. It is not enough to know the structure of the molecules with hormonal activity; we need to know as well the structures of the molecules on which the hormones act before we may begin to understand the relationship between chemical structure and biological function.

The slight variability of the amino acids in the insulins of different species of mammals is interesting from the point of view of evolution. Nature paints the image of life with bold, uniform strokes; however, the loose bristles on the edge of her brush do often trace minor patterns distinctive of each species.

No mammal has survived a mutational change involving the total loss or even gross changes in the structure of insulin.

The minor changes which nature has permitted at the molecular level might form the basis of a new system of taxonomy: "The horse and the whale are two species of mammals which differ in amino acid 39 in the A chain of insulin."

Paralleling Dr. Sanger's work on the structure of insulin another monument of architectural biochemistry was being shaped by Dr. Vincent du Vigneaud of Cornell University.

Dr. du Vigneaud, our foremost expert in the biochemistry of sulfur compounds, cut his scientific eye teeth, interestingly enough, on the chemistry of insulin. Then, almost thirty years ago, his attention was attracted to the hormones of the posterior lobe of the pituitary gland, because these hormones, like insulin, were also known to be rich in sulfur compounds.

Since the work on the hormones of this gland is repetitive, let us concentrate on just one hormone. Oxytocin has profound physiological effects on the pregnant human female. Minute amounts of it can induce labor by causing uterine contractions, and it can also induce the flow of milk.

The isolation of the hormone was one of those Herculean feats lasting over decades. It entailed the extraction of tens of thousands of beef glands, the application of new tools of purification as they were invented, and, finally, the patient cajoling out of the pure product from the last traces of contaminants.

Dr. du Vigneaud analyzed the pure hormone and found that it contained eight amino acids, each of which appeared only once in the molecule. This was fortunate because such a structure is relatively simple. Simple, however, only in comparison to the structures of proteins; for oxytocin, even with only eight amino acids, can exist in 8! or 40,320 different forms.

The structure of the hormone was determined by techniques essentially the same as those described under the work on insulin. The first amino acid was determined with the use of Sanger's reagent, the various amino acid sequences were deduced by the partial fragmentation of the molecule and identification of the amino acid complexes by paper chromatography.

From this pooled information Dr. du Vigneaud decoded the sequence of the eight amino acids. The relatively simple structure of oxytocin offered to Dr. du Vigneaud the tantalizing possibility of its synthesis.

Dr. du Vigneaud came to this final assault, the synthesis of the most complex biological material ever attempted, with unique qualifications. He had previously demonstrated rare genius in structural biochemistry: he had synthesized the vitamin biotin, and he was the first to synthesize penicillin. Moreover, his scientific ability is matched by a rare skill at organization. In a lecture summarizing his work on the hormones of the posterior pituitary gland he listed twenty-five scientific associates who had collaborated with him in various segments of the total task.

It would require a whole book to describe the logic and the techniques of organic chemistry on which the planning and execution of the synthesis of oxytocin was based. Synthetic organic chemistry is one of the most extraordinary achievements of the human mind. It has been built on a curious mixture of rather messy and pedestrian experimental techniques and an ingenious system of logic which uses crude symbolic images to describe a variety of attributes of the compounds of carbon. The images depict both structure and function. The validity of this system of logic has passed untold pragmatic tests: hundreds of products of the living cell have been duplicated in the flasks of the organic chemist. One of the most defiant challenges to this system of inference was the structure and synthesis of oxytocin. Does this complex man-made hormone have any physiological potency on man? Dr. du Vigneaud handed over to clinical associates his product to be compared with natural oxytocin. "They were found to be indistinguishable in effectiveness. Approximately a millionth of a gram of either the synthetic or natural material given intravenously to recently parturient women induced milk ejection in 20 to 30 seconds."

The synthetic oxytocin is now available commercially and is used by obstetricians for the induction of labor in the terminal stages of pregnancy. Not the least valuable asset of this drug is our ability to time with its aid the onset of parturition at the convenience of patient and physician. Thus a product shaped by the hands of the chemist may produce a technological revolution in obstetrics: it will no longer be a nocturnal profession.

The biological activity of the synthetic material proving it to be identical with the natural was the laurel wreath on twenty-two years of inspired effort.

The Nobel Prize Committee bestowed its accolade upon both Dr. Sanger and Dr. du Vigneaud. Their achievement equals, many of us think, the stunning accomplishment of the atomic physicists. These triumphs—of the biochemist and the physicist—are the cumulative products of different disciplines, indeed of different types of human minds. Structural chemistry is based on a system of inference rooted in images; the discipline which opened the core of the atom has a mathematical foundation. Since the skill of the physicist struck terror in our hearts, those who know about it outnumber those who understand it by legion, but the equally spectacular and potentially more beneficial achievement of the biochemist is known only to the handful who understand it.

Blood is a truly remarkable juice.

8 . BLOOD

THE HIGHWAY TO OUR CELLS

"BLOOD IS A TRULY REMARKABLE JUICE," said Mephistopheles to Dr. Faust as the two went through the sealing of their contract.

Mephistopheles showed rare biochemical insight: blood *is* a truly remarkable juice. It is a juice to which we owe much. We owe to it our size, we owe to it our brain, we owe to it our wonderfully complex physiological existence.

It was the great French physiologist, Claude Bernard, who pointed out, in 1878, that the evolution of the highest forms of life has been made possible by the liquid *milieu intérieur*. "The living organism," he wrote, "does not really exist in the *milieu extérieur* [the atmosphere for terrestrial animals; salt, or fresh water for those who had not invented lungs] but in the liquid *milieu intérieur* formed by the circulating organic liquid which surrounds and bathes all the tissue elements."

Complex life is possible for the biological organism only with adequate means of transportation from organ to organ, just as complex social life is made possible only by transportation facilities from community to community.

Blood is the highway to the cells of our tissues. Without it the cells, even those on the surface of our skin would perish.

They would lack oxygen; they would lack food; they would be killed by poisons of their own making.

The center of this most wonderful system of transportation is, of course, the heart. It pumps, in a lifetime of seventy years, about two billion times, and it pushes on its path a hundred million gallons of blood. From the right heart to the lungs, from the lungs to the left heart, from there, through the arteries into the capillaries of the tissues, back through the veins into the right heart, 'round and 'round goes the blood in its wondrous, uninterrupted circle, performing many chores on its rounds. It takes carbon dioxide from the cells and exchanges it for oxygen in the lungs. It is a traveling department store of foodstuffs. It carries everything a cell needs: amino acids, fats, sugars, vitamins, and salts. A hungry cell in one of the outlying districts, say in the toe, extracts from the blood swishing by whatever it requires: a few million molecules of glucose, a hundred thousand molecules of vitamin B_6, a few thousand cobalt atoms. Each cell, however far removed, is thus as well provisioned as a cell in the heart itself.

In addition to foods, the blood carries a variety of other wares: hormones to stimulate laggard cellular mechanisms; antibodies to battle invading poisons; clotting agents to seal breaches in its cyclic path. Furthermore, the blood distributes the heat evolved from the furnace of the cell, thus maintaining a uniform temperature throughout the body.

Finally, the acidity of tissues is kept within tolerable bounds by the blood. Life is fenced in within very narrow limits of acidity. A variety of acids are produced by the metabolic activities within the body cells. Carbon dioxide, lactic acid, and uric acid all tend to acidify the human body. Excess acidity slows down many enzymes, and unless the acidity is counteracted, life itself slows down and eventually halts. Many bacteria

—for example, those which turn milk sour—are destroyed by their own metabolism. They produce lactic acid and throw it out into the surrounding fluid. Soon, so much acid accumulates that they become the victims of their own sewage. The blood contains powerful neutralizing mechanisms ready to pounce on the acids cast off by the cells.

How does this "remarkable juice" perform its many functions? It consists of both cells and a variety of noncellular, dissolved materials. About 45 percent of the blood is composed of the red cells. They are tiny red discs (five million of them are packed into a volume the size of a sugar crystal) containing the red protein hemoglobin. This is the truck on which oxygen and carbon dioxide are shuttled back and forth. It is an unusual truck, the hemoglobin molecule, it can carry only one five-hundredth of its weight of oxygen.

A hemoglobin molecule is made of a colorless protein, globin, and an iron-containing pigment, heme. Oxygen unites with this complex molecule in the lungs forming a definite chemical combination which, however, is easily decomposed in the capillaries, liberating oxygen to the gasping tissue cells. Once freed of the oxygen, the hemoglobin forms a temporary chemical alliance with carbon dioxide which is then ferried to the lungs.

Unfortunately, hemoglobin is not very discriminating as it forms its chemical alliances. Life can be snuffed out by carbon monoxide poisoning because of this lack of discrimination by the hemoglobin. It combines with carbon monoxide gas with far greater avidity (two hundred times greater) than with oxygen. Thus, if one inhales a mixture of oxygen and carbon monoxide, the two gasses compete for the favors of the hemoglobin, and oxygen has odds of two hundred to one against it in the contest.

Death from carbon monoxide asphyxiation is caused simply by the lack of free hemoglobin to transport oxygen to the starving tissue cells. The hemoglobin-carbon monoxide combination is actually not poisonous. City dwellers invariably have about one percent of their hemoglobin tied down to carbon monoxide, and tobacco smoking immobilizes as much as five percent of the hemoglobin. (Carbon monoxide is produced by the incomplete burning of the tobacco.) Fortunately, the combination between hemoglobin and carbon monoxide is not permanent. The gas is quickly swept out of the system, unless, of course, enough was absorbed to overwhelm the victim.

Red cells are highly specialized for their role as trucks for the transport of oxygen and carbon dioxide. Their metabolism is very low; they are stripped down to such an extent that the mature cells lack even a cell nucleus. They are nourished by the metabolism of other cells as long as they are useful, but when in old age they falter at their tasks they are liquidated. There are special cells in the blood vessels of the spleen and liver which unceremoniously devour the aging red cells. These cannibalistic phagocytes (eating cells) are constantly trapping and dismembering the more sluggishly moving red cells. The iron is salvaged from the wreckage but the pigment, heme, is piped into the gall bladder, from where it is discharged into the intestine. What happens to the protein, the globin, we do not know.

Why the red cells must die is but one of the unsolved mysteries connected with them. There are many others. How do the phagocytes select their aging prey for the slaughter? Are there stigmata of age, or do the phagocytes merely fall upon the laggard cells?

All cells, young and old, travel under the impetus of the pumping heart. What makes an aging cell slow down in the capillaries? Is there a definite retirement age for red cells or are they destroyed at random?

Only to the last question do we have an unequivocal answer. The meteoric existence of the red cells was known for a long time, but estimates of their career varied from five days to two hundred days. Recently the life span of the red cell was clocked with isotopes. The counting of the days of the red cell was an unexpected by-product of an entirely different project—as so often happens in scientific research.

A biochemist who completes merely his original projects is rather limited. The chemical ways of the cell are so much more complex than we can at present imagine, that, during the course of almost any project, mechanisms of far greater interest than the one originally visualized are invariably exposed to observant eyes. This is why freedom for the investigator to follow up unexpected, chance findings is so essential.

The metabolism of the simplest of amino acids, glycine, was to be studied in humans. Dr. David Shemin at Columbia University made two ounces of glycine which contained not the ordinary isotope of nitrogen but its rare, heavy isotope, and, showing the ultimate in confidence in the purity of his preparation, he promptly ate it. For weeks thereafter he obtained samples of his own blood and measured the heavy nitrogen isotope in the various fractions of it. The largest concentration of the isotope was in the red pigment of the hemoglobin. Apparently this pigment, heme, is fashioned out of the glycine molecule, hence the high concentration of the heavy isotope in it. Glycine which contains but one nitrogen and two carbon atoms can be marshaled by the enzymes of the bone marrow, the birthplace of the red cells, into the elaborate structure of heme which contains thirty-four carbon and four nitrogen atoms. The carbon atoms do not all originate from glycine. A four-carbon component of the sugar metabolizing Krebs cycle first fuses with glycine and then this newly formed thread is woven into the tapestry of the hemin molecule. This is one more example of the great versatility of the cell in fashioning its components of intricate structure and wondrous variety from abundant small molecules.

The entry of glycine into the hemin molecule pointed a way for the measuring of the life span of the red cell. Only those red cells which are made on the day or two following a meal of labeled glycine contain the heavy isotope of nitrogen. As these cells are devoured at the end of their careers, the isotope-containing heme from them is voided through the bile and lost from the body forever. Therefore the disappearance of the heavy isotope from the heme marks the death and disposal of the cells produced on the day of the isotopic meal.

Small samples of blood were tapped almost daily after a meal of labeled glycine. The amount of heavy nitrogen remained at a constant level for about eighty days. After that it began to disappear. From a mathematical analysis of the complete data we can calculate that the average life span of the human red cell is one hundred and twenty days. The "average" is emphasized, for there was a small amount of heavy nitrogen isotope left even after one hundred and thirty days. Apparently red cells, just like the whole organisms of which they are a peripatetic part, vary in their life span. Some red cells exceed, and some fall short of the six score days.

The rate of destruction and of synthesis of red cells is prodigious: about ten million red cells are born and about the same number die every second in each of us. The number of red cells can increase for a variety of reasons. One of the most interesting of these is prolonged stay at high altitudes. Since the concentration of oxygen at high altitudes is low, the body adapts itself to the emergency by making more cells and hemoglobin for the transport of the gas in short supply.

The ability to adapt to lowered oxygen concentrations is a valuable asset. For the greater the adaptability to changing environment, the greater are the chances of survival of an organism. The red blood corpuscles are not the only cells which can be augmented to compensate for an environmental or structural deficiency. If one kidney is removed from an animal the other kidney will increase in size, thus enlarging its functional surface and enabling it to carry the larger burden which falls upon it. The mechanism of the induction of the formation of new cells by a chemical stress—a decrease in the available oxygen or an increase in the amount of waste products awaiting disposal—is a challenging problem for the biochemist. For here he is coming to grips with the basic unsolved problem: the method of synthesis of a new cell, complete with its structural components, its enzymes, and its urgency to live. We are just beginning to explore the mechanism of adaptation with the tools of biochemistry.

If the red cells are abnormally low in number or are deficient in hemoglobin, one is said to be anemic. Excessive hemorrhage is the simplest and most easily remedied cause of anemia. Fully one fourth of an animal's blood can be lost with impunity: the loss is made up in two to three weeks. This is the reason for the ease with which we can donate a pint of our blood, which is but one tenth of our total wealth of it. Another cause of anemia, particularly among infants, is lack of iron in the diet. Still other causes are poisons which destroy the bone marrow. Finally, the anemia which at one time was as dreaded as cancer is today is pernicious anemia. A little over thirty years ago this disease was as relentlessly fatal as an advanced case of inoperable cancer is today. But the disease is now mastered. Its conquest is a monument to the joint efforts of medicine and chemistry.

Until 1926 we had no idea of the cause of the disease. Infection, poisons, and cancer were all accused as the possible culprits. The methods of treatment were as varied as they were futile. There was an odd brake on progress against the disease: no experimental animals could be induced to come down with it. We can make dogs diabetic by performing a simple operation; we can transplant tumors; any of the infectious diseases can be implanted into almost any animal; but to pernicious anemia all but the human animal seemed to be immune.

Dr. George H. Whipple and his associates at the University of Rochester decided to study experimentally produced anemia even though it did not resemble pernicious anemia. Dogs were bled copiously and frequently to induce this simple anemia. The goal of the study was to see if we could intercede by some means and accelerate the rate of regeneration of red cells. Various dietary aids were tried and it was found that feeding beef liver to anemic dogs helped their recovery.

After this discovery had been made, the next obvious step was taken by two physicians at Harvard, George R. Minot and William P. Murphy. They fed to their patients suffering from pernicious anemia huge amounts of liver and noted marked improvement within ten days.

As long as the patients were kept on a diet of about a pound of liver a day their improvement continued. Apparently there was something in liver which could protect a patient against the ravages of pernicious anemia.

That is as far as the physicians could go in the search for the cure. At this stage the biochemists took over and began to track down with their specialized searching tools the active principle in the liver. The approach of the chemist to such a problem is, by now, familiar to the reader. Using a variety of chemical manipulations—extraction, precipitation, evaporation —the chemist weeds out the inactive contaminants, testing at each stage the efficiency of his gardening by an assay of the potency of his preparations. Needless to say, the product is expected to become more and more active: a smaller and smaller weight of it should contain most of the original biological activity. Crystallized enzymes, vitamins, hormones, and essential amino acids are the testimony to the effectiveness of these methods of the chemist.

This search, however, was hindered by a particularly difficult obstacle. The only test for the presence of the active principle in the liver was the improvement it brought to human patients suffering from pernicious anemia and, with the easily administered liver therapy, the untreated disease was becoming more and more rare. One leading research center offered free hospitalization and medical care to any patient suffering from the disease in return for permission to standardize the preparations with which the patient was being cured.

Only twenty years after the search began was this particular obstacle removed. During those twenty years chemists managed, despite the lack of convenient testing, to concentrate the active principle several thousandfold. Whereas a pound of liver had to be fed to a patient each day, only one milligram (450,000 milligrams make a pound) of the concentrated material was needed per day when it was given in the form of an injection.

The outstanding biochemist engaged in the purification of the liver factor was the late Dr. Henry D. Dakin, one of the stalwart pioneers of American biochemistry. (The antiseptic solution which he devised, and which bears his name, was the life-saving antiseptic of the First World War.)

A point was reached in the purification of the liver factor beyond which further progress seemed almost impossible. The product, even though enormously concentrated, was still far from pure. Testing the material became well nigh impossible: response by the rare human patients to the various concentrated preparations was almost uniform.

Then, in 1946, a brand new kind of dietary deficiency was reported in the rat. If rats were kept on a diet in which the protein source was alcohol-extracted casein, they failed to grow. (Of course the diet was complete in all the known vitamins.) The missing factor could be found in a variety of foods. It was also present in the commercial liver extract which was being used to cure human pernicious anemia. Let us call this unknown material the "rat-growth factor."

There had been studies on the nutrition of chickens, dating back twenty years, which showed that unless laying hens were fed some protein of animal origin, their eggs did not hatch normally. The factor present in animal proteins and essential for normal hatching was called the "animal-protein factor."

There were then three different unknown dietary factors: the pernicious-anemia factor, the rat-growth factor, and the animal-protein factor. These three apparently different problems were tied together and the three factors were proved to be the same, as soon as a pure crystalline material was isolated. Before that could be accomplished, however, a rapid, consistent, and specific assay for these factors was needed. The assay for one of the three apparently divergent factors was provided by Dr. Mary Shorb. She found a microorganism—*Lactobacillus lactis* Dorner, or LLD for short—which requires for *its* growth the "rat-growth factor."

The use of microorganisms for assay purposes is a recent development of immense value. We have seen earlier that

yeasts and animals need the same vitamins, or that the amino
acid methionine is essential in the diet not only of men but
of mice and of microorganisms as well. Among the thousands
of different microorganisms we can find some which require
for their growth any of the dietary essentials of animals. Using
them instead of animals for assay purposes is a great advantage.

In order to deplete their reserve of a dietary essential, animals sometimes have to be kept on a deficient diet for months.
Indeed a deficiency often does not become apparent until the
second generation. It takes months to raise a generation of rats
but only minutes to raise a generation of microorganisms, so
rapidly do they reproduce. Therefore, assays with microorganisms are a matter of hours.

The principle of such an assay is simple. The growth of bacteria is proportional to the amount of balanced diet available
to them. If their diet contains all the essentials but one, that
one factor, whether it is a vitamin, or an amino acid or a salt,
becomes the limiting factor in the growth of the organism.
Growth becomes proportional to the amount of the limiting
factor available. For example, if a million bacteria grow on one
microgram (one twenty-eight-millionth of an ounce) of vitamin
B_x, two million will grow on two micrograms of that vitamin.
If now, on an unknown amount of vitamin B_x 1.5 million bacteria will grow, we can then conclude, that the unknown sample
contains 1.5 micrograms of the vitamin.

Dr. Shorb found a microorganism—LLD—which needed
something for growth besides the then known dietary factors.
Her LLD factor was found in the same foods which contained
the rat-growth factor. Liver extracts were the best source of
the LLD factor. Dr. Shorb suggested that perhaps the LLD
factor and the pernicious-anemia factor were the same. If this
were so, then here, at long last, would be an assay for the

pernicious-anemia factor other than a test on a human patient.

Within a year chemists at Merck and Company isolated the pernicious-anemia factor in pure crystalline form. It was the LLD factor, the rat-growth factor, and the animal-protein factor as well. Since the product was now a definite, pure compound and not just a vague "factor" it was entitled to a new name. Vitamin B_{12} was the name given to the shiny red needles which cure pernicious anemia, enable LLD to grow, permit normal hatching of hens' eggs, and allow rats to grow normally.

Vitamin B_{12} is one of the most potent of the known biologically active materials. It is effective in even smaller amounts than is biotin. The late Dr. Randolph West of Columbia University, one of the foremost clinical experts on pernicious anemia, who was Dr. Dakin's clinical collaborator in the early days of the search for the anti-pernicious anemia factor, was, very appropriately, the first to report on the clinical potency of vitamin B_{12}. He found that as little as 3 micrograms, or one ten-millionth of an ounce, when injected into a human, started an immediate improvement in the patient's blood picture.

The structure of vitamin B_{12} has been decoded. A remarkable constituent of it is the metal cobalt. That traces of cobalt are essential in the diet of the mammal was previously known. There are a number of such trace elements essential for health. Indeed, the list of elements essential in our nutrition begins to read almost like the chemist's Periodic Table of the elements. Now the specific role of one of them, cobalt, became clear. It is part of vitamin B_{12}.

While knowledge of the functions of the inorganic salts is accumulating but slowly, it is known quite well what elements are present in blood and in what quantities. (It is easier, by far, to assay for an element than to pin down its biological function.) Blood contains the same salts as sea water, and these salts are present in approximately the same ratio. However, blood is fivefold more dilute. In other words, the concentration of each element present in blood is one fifth that in sea water.

That their blood is but diluted sea water—with respect to inorganic salts—is telling circumstantial evidence for the marine origin of animals. The odds are enormously against the chance repetition of the same ratio of the same elements in the sea and in blood. The dilution of the blood of animals has been explained through studies made not by biologists but by oceanographers. Apparently, it is not blood which has become diluted but rather it is the ocean which has become more concentrated since animals arose from that vast aqueous cradle. It is known that the inorganic salt content of the ocean keeps increasing steadily. There have been some interesting extrapolations from the rate of that increase to estimate the era when the ocean had one fifth of its present concentration. It is said that the calculated time coincides with the time animals are estimated, from other lines of evidence, to have evolved from their marine precursors.

In addition to the inorganic salts, blood contains dissolved gases. Of these nitrogen is the most abundant and the least useful. (Oxygen and carbon dioxide are not merely dissolved; they are held in chemical bondage.) Nitrogen seems to have no function other than that of a diluent for oxygen. In the gaseous elementary form nitrogen is completely unusable and can be replaced by another gas, for example helium, in the gaseous mixture that an animal breathes. Indeed, that is precisely what is done with the gases pumped down to deep-sea divers in order to minimize the possibility of their experiencing caisson disease, or the bends. This is a neat application of some simple laws of physics to eliminate a dangerous occupational hazard. Caisson disease is caused by the greater solubility of nitrogen in blood at higher pressures. If the pressure on a diver is quickly released—by too rapid surfacing—the nitrogen dissolved during his stay in the depths is suddenly released, forming bubbles in his blood vessels. At that moment the diver's blood simulates in appearance the contents of a bottle of carbonated cherry beverage from which the sealing cap is suddenly removed. The release of gas bubbles in the veins and arteries causes violent pain, convulsions, and, in severe cases, death. Helium is quite as inert as nitrogen in our bodies but is far less soluble. Thus when inhaled even at great pressures it does not endanger the diver, for too little of the gas dissolves to liberate bubbles at atmospheric pressure.

The freedom of the whale from the bends—this occupational hazard of other deep-sea divers—has puzzled biologists for several generations. According to whalers, a wounded whale can take a half a mile of line to the depths and can then surface with startling speed. An equally rapid ascent from two hundred to three hundred feet would surely release enough bubbles of nitrogen to kill a man. How does the whale tolerate the tremendously greater changes in pressure?

It has been claimed that the whale harbors in its blood stream myriads of a certain species of microorganisms which sop up the inhaled nitrogen of their leviathan host. These microorganisms, like those in the nodules of the clover, were supposed to be nitrogen fixing; such organisms have the ability to convert gaseous, molecular nitrogen into soluble compounds of it. The interpretation offered by the discoverer of nitrogen-fixing bacteria in the whale was that as fast as the whale absorbs nitrogen from its lungs the gas is sponged up by the enzymes of the tiny inhabitants in its blood stream. This would be a fascinating teleological symbiosis, and it ought to be true—but it isn't. The observation could not be repeated. The organisms found by

the original investigator must have been post-mortem contaminants. This episode reaffirms two truisms: Scientists are
fallible; Moby Dick is not: She still keeps her mysteries well
hidden.

The noncellular part of the blood, the plasma, contains 7
percent dissolved proteins. This, in addition to the 14 percent
of hemoglobin, brings the total protein content of blood to 21
percent. That the hemoglobin is not freely dissolved but is
packaged in the red cells is one more example of the remarkable chemical foresight of nature. A 21 percent protein solution
would have tremendous viscosity and flow would be impossible.
Blood would indeed be "thicker than water." As it is, the hemoglobin is safely wrapped up in the red cells and does not im-
pede the easy flow of blood.

The free plasma proteins contain the clotting mechanisms
and the antibodies. Great strides have been made, particularly
since the start of the Second World War, in the separation of
plasma proteins into their various fractions. The products have
widespread clinical uses. Absorbable surgical threads and fibrous
packing which stop hemorrhage during surgery are everyday
tools of the surgeon, made from human plasma proteins.

These ancillary aids to medicine are but some of the sweet
fruits which grew from the evil garden of war. Another unexpected fruit is an understanding and, one hopes, the possible
control of one of the diseases of blood, sickle-cell anemia. Sickle-
cell anemia is a hereditary disease which apparently afflicts only
Negroes. The symptom which distinguishes it from other types
of anemia is the presence of abnormally shaped red cells in the
veins of the patient: some of the corpuscles are crumpled into
crescent or sickle shapes. About 8 percent of our Negro population have some sickle cells in their veins. Fortunately only
about 2 or 3 percent have the abnormal cells in sufficiently large

numbers to incapacitate them. The puzzling feature of the disease until recently was that the corpuscles have their normal disc shape while they are in the arteries but crumple into shapeless bags in the veins. The disease, therefore, is not caused merely by a malformation of the cell membranes but rather must be due to an abnormal response either by the membrane or its contents to changes in the concentration of oxygen or of carbon dioxide as the corpuscle exchanges these gases in the lungs.

Dr. Linus Pauling is a physical chemist who became interested in biological problems several years ago. He heard of sickle-cell anemia for the first time while he was serving on a committee appointed by President Roosevelt to study means of advancing medicine. Dr. Pauling very naturally approached the disease as a physicochemical problem. He and his associates isolated hemoglobin from the blood of patients with sickle-cell anemia. They then extracted from the hemoglobin the red pigment, heme, in a pure form and compared it with heme obtained from normal humans. The hemes from the two sources were exactly the same. The protein fraction, the globin, was next purified and was made the target of the battery of critical tests and measurements which reveal differences in proteins. One such test is the measurement of the rate of migration of a protein toward one of the poles in an electrical field. Proteins have the ability to assume either a positive or a negative surface charge depending upon the acidity of the liquid in which they are suspended. (Their ambivalence has earned them the name "zwitter-ions," meaning hermaphroditic charged particles.) The electrical charges on the surface of the protein molecule will determine the direction of migration in an electrical field. A level of acidity was reached at which the globin from normal red cells moved toward the positive pole and the globin from sickling cells moved toward the negative pole. In other words, the protein from the diseased cells seems to carry a larger positive charge than the normal protein does. Such an abnormal charge can have profound effects on the protein's ability to bind water or the gases carbon dioxide and oxygen to it, and, also, on the ability of the protein molecules themselves to be bound together. Pauling's discovery revealed the cause of the sickling phenomenon: when the abnormal protein is laden with carbon dioxide in the veins it shrivels in volume and fails to fill out the cell membrane. His finding also provides a starting point for an approach to a possible remedy of the disease. An attempt to alter the properties of the abnormal protein by means of chemical agents is under way.

Pauling called sickle-cell anemia a "molecular disease." While the name is effective in calling attention to the necessity and fruitfulness of studying biological mechanisms at the molecular level, it is somewhat misleading. All the metabolic diseases and probably all of the degenerative diseases are "molecular." In the case of sickle-cell anemia we happen to know the specific molecule which is "diseased." Other than that the disease is not unique and does not merit a special, generic name.

We even know why the protein is abnormal, why it is deficient in negative charge and is consequently more positively charged. The protein of hemoglobin contains nineteen different amino acids repeated enough times to yield an aggregate of about six hundred amino acids. One of Dr. Pauling's associates traveled to Dr. Sanger's laboratory at Cambridge University to explore with Sanger's tools of analysis the hemoglobin obtained from patients with sickle-cell anemia. The task appeared to be formidable. The protein of hemoglobin is ten times as large as insulin; thus the decoding of the complete structure might take a decade or more. Therefore a bold gamble was decided on. The protein was broken down by the enzyme trypsin to fragments containing from seven to ten amino acid units. The hemoglobin from normal human blood was similarly crumbled down and the fragments from the normal and sickle-cell anemic blood were separated and compared. A fragment was found in normal hemoglobin in which glutamic acid, an amino acid, appears twice in sequence. In the abnormal protein from sickle-cell anemia these twin glutamic acids were absent: one of the glutamic acids was replaced by another amino acid, valine. Glutamic acid is one of two unique amino acids which have an extra acidic group in their structure. Such acidic groups confer a negative charge on a protein. The source of the deficiency of the negative charge on the abnormal protein molecules of sickling cells was now pinpointed: they lack a glutamic acid. A crippling and often fatal disease is caused by one slip as the amino acids are woven into a strand of hemoglobin. In some ancestor of the people afflicted with sickle-cell anemia an unfortunate mutation had taken place, producing a gene which blunders in one tiny area of the pattern as it weaves the tapestry of the hemoglobin molecule.

That such a fumbling can occur is a source of terror; that it occurs so infrequently is a source of awe. The precision required for the perpetuation of life is almost beyond belief. A human being has thousands of different species of protein molecules. Each one of these has a unique structure; each has an amino acid sequence all its own. Billions of each of these protein molecules must be made every second of our life to shape and to sustain it. The pattern of each protein must be unerringly duplicated. If there is but one fumble and just one amino acid is misplaced, as in sickle-cell anemia, the red cells crumple shapelessly, causing disease or death. This one blunder with its patently gruesome consequence testifies to the unfailing success in the shaping of thousands of other kinds of protein molecules in our bodies. Disease or health, indeed, life or death, hangs on a thin thread no stronger than the link between two appropriate amino acids.

> *"We shall fight on the beaches, we shall fight on the landing grounds, we shall fight in the fields and in the streets, we shall fight in the hills; we shall never surrender."*   CHURCHILL

9. CELL DEFENSE

MANY are the foes which attack the cell. The assaulting hordes come in strange shapes and varied sizes and aim a diverse battery of weapons at their target. The cell fights back with sustained valor. If its armed vigilance falters, disaster befalls the cell fortress. The struggle starts at birth and continues relentlessly. There is no quarter, no armistice; only survival or death. That cells do survive is a miracle wrought by their defensive weapons—weapons of great variety and of astonishingly ingenious design.

What particular cell-guard weapons from the well-stocked arsenal are mobilized at any one time depends upon the nature of the invader. The cell can burn up the invading enemy; it can, by means of enzymes, alter the marauder to remove its sting; it can fashion special shock troops which, using their strands of protein molecules grapple with the invader until it is immobilized.

Let us observe some of the embattled units in this constant warfare.

The best defense against any foe is to prevent its penetration. The body's first line of defense is a tough skin, which, though quite effective, has, unfortunately, some weak spots. Bacteria lodge in the pores of the sweat glands and in the hair pits, causing pimples and boils. Poisons and bacteria can pour through the larger openings, the mouth, nose, and eyes. The stomach, however, is booby trapped against the bacteria: the high concentration of acid in its juices kills most of them.

If a simple, chemical poison enters the body it is handled with effective vigor. The defensive campaign follows a well-defined strategy. The largest route of entry for such poisons is, of course, the mouth. The first maxim of the strategy is: Absorb as little of it as possible. A poison can do us no harm while it is in the alimentary canal. The transient contents of that long tube are actually not part of the body; they can do us good or ill only when they gain admission into the blood or other tissues.

There is a remarkable screening performed during absorption. Before a substance can gain admission into the tissues it must pass the discriminating scrutiny of the cells lining the alimentary canal, and these are remarkably adept at excluding undesirables. For example, humans absorb cholesterol with ease. However, there is an astonishingly similar substance made by plants, sitosterol, which is not absorbed at all.

If a poison passes the selective barrier of the alimentary canal, the appropriate order of the body's high command is: Excrete it! Use the kidney, use the lungs or sweat glands, but excrete it. This is, of course, a selectivity too, but in reverse. Now the foreign substances are preferentially expelled.

If excretion fails, the command is: Burn it up! To study the fate of toxic substances in animals we inject them. In this way we by-pass the forbidding scrutiny of the alimentary canal. If we insert into the muscles of an animal some benzene, most of it is promptly oxidized (mostly to carbon dioxide and water), by the enzymes of the victim, and the animal is rid of the poison.

The enzyme systems which burn up toxic substances are not teleologically designed for this single, self-defensive purpose. These enzymes are always present and are usually performing their normal metabolic oxidations. But if a toxic, foreign substance comes along and happens to fit into the working pattern of an enzyme system, the animal benefits from the versatility of its enzymes.

It would be difficult to visualize how an organism could be equipped to handle any poison it might encounter with enzymes tailored a priori to that purpose. There was a period in the development of biochemistry when hundreds of different organic compounds were administered to animals in order to study their fate in the body. Many of these substances had undoubtedly never existed in the universe until zealous organic chemists strung them together. But still, even though the animal had never, in its whole evolutionary history, encountered these substances, it promptly oxidized them or made extensive alterations in their structure.

During the Second World War there was information that the Germans were manufacturing on a large scale a substance called diisopropylfluoro-phosphate. (Even chemists call it DFP.) Since DFP paralyzes its victims, it was feared that the Nazis might use it as a "nerve gas." DFP was therefore extensively studied by the medical unit of the Chemical Warfare Service. Dr. A. Mazur, a biochemist of that service, found an enzyme in the liver of the rabbit which tears DFP apart. Now, DFP does not exist in nature and undoubtedly it never has. If it were not for Hitler, no rabbit might ever have made the unpleasant acquaintance of DFP. Yet rabbits have the enzymes to dismember this rare poison.

There is other evidence that these so-called detoxication processes are performed at random. A poison sometimes be-

comes *more* toxic after the enzymes finish their alterations of
it. The fate of ethylene glycol, the poisonous solvent for sulfanilamide (see page 76) is a good example of this.

Ethylene glycol (the reader may know it as an antifreeze),
is an alcohol. Many alcohols are poisonous—even ethyl alcohol,
which is the least poisonous of them, can be quite dangerous.
Let us digress a bit from the fate of ethylene glycol and follow
the course of ethyl alcohol in the body.

Ethyl alcohol can reach fatal concentrations in the body
from the rapid intake of about one tenth of an ounce of 200-
proof alcohol per pound of body weight. In other words, if
a man weighing 150 pounds drinks rapidly 15 ounces of 200-
proof alcohol, his chances of recovery from his alcoholic stupor
are mighty slim. (A variety of factors such as the state of the
subject's health and his previous experience in the consumption of alcohol make the exact outcome of the experiment unpredictable.)

Alcohols produce their toxic effects by inhibiting the rate
of respiration within the cells. The assortment of symptoms
which mark a man as being intoxicated can be induced in the
most righteous teetotaler, without the use of a drop of alcohol,
simply by reducing the concentration of oxygen in the atmosphere he breathes. The scarcity of oxygen will limit cellular
respiration and the external symptoms of reduced respiration
are the same whether it is brought about by an inhibitor or by
diminished supplies of oxygen.

The influence of low oxygen concentration has been studied
extensively since it is an important factor in the efficiency of
pilots—unless, of course, they breathe bottled oxygen. (At
12,000 feet the oxygen in the atmosphere is one third of what
it is at sea level.) The pioneer in these studies, the English
physiologist Joseph Barcroft, reported that on journeys to high

altitudes he has witnessed emotional reactions similar to those experienced after an overdose of alcohol: depression, apathy and drowsiness or excitement and joyfulness, and general loss of self-control. "A person may sing or burst into tears for no apparent reason or be extremely quarrelsome, indolent, and reckless." During the Second World War, some members of the crews of high-flying bombers would take off their oxygen masks on the return trip, for a quick, hangover-less bender.

The fate of all alcohols is the same in the body. They are gradually oxidized. In the case of ethyl alcohol the intermediate stage during the course of the oxidation is the formation of acetic acid, a normal constituent of the body. Acetic acid can be either oxidized further to carbon dioxide and water or utilized as a brick for the assembly of a number of more complex body components. Alcohol, therefore, is really a food. A very limited food, to be sure, since, like sugar, it lacks proteins, vitamins, and minerals.

Now, just as ethyl alcohol is first oxidized to an acid, so ethylene glycol is also oxidized to an acid. But this acid, oxalic, happens to be a merciless poison. The enzymes which accomplish this oxidation doom the animal to quick death. In this case it would be much better for those enzymes to lie low and do nothing to the ethylene glycol. The animal might be able to excrete it slowly through the lungs and kidneys and thus, if the dose is not too large, survive. Oxidation of even much smaller doses brings certain death.

The struggle against our bacterial enemies seems to be more purposeful than the disposal of simple toxic substances. However, even this apparently planned campaign may be really more haphazard than it appears. Bacterial poisons are either proteins or other complex molecules. We are unable, with our present knowledge, to view the intimate mechanisms that accomplish

the disposal of these poisons. We only see the ultimate effect
and that seems mighty purposeful.

Those parts of the body which are not normally in contact
with matter from the outside world—the blood and various
other tissues—are free from bacteria. Not statically, the way
the inside of a can of evaporated milk is bacteria-free, but dy-
namically free. If any bacteria penetrate into the inner tissues
they are attacked by special shock troops for aggressive defense
—the white cells of the blood. They are smaller in number than
the red cells and, as their name implies, they are without the
red hemoglobin. They have, instead, other specialized equipment—enzymes. The white cells can project tiny strands of
their tissues and encircle the bacteria. Once trapped the bacteria are helpless, for the potent enzymes oozing out of the white
cells tear them to shreds.

A boil on the neck is a typical battleground in such a struggle
against marauding enemies. Some microorganisms lodge near
the root of a hair and, finding a warm, cozy nook and a source
of food, begin to multiply. As a first reaction to this breach
in the body's static defense, the area becomes inflamed. The
capillaries become dilated, causing the flow of an augmented
supply of blood. Fluids from the blood escape into the area
and form a clot, converting the whole into a jellylike mass. A
barrier of fibrous tissue is formed around this mass, isolating
the infection. Meanwhile the white cells have been gathering
on the battleground and, crawling through the capillary walls,
they pounce upon the infecting organisms. If all goes well, the
invaders are killed off in this melee. If not, the infection spreads
and the battle is repeated at every new focus of infection. The
white cells carry out Mr. Churchill's magnificently expressed
strategy: wherever the germs may travel in the blood stream,
they are pounced upon and engaged in mortal combat. The white

cells get powerful assistance, once the infection becomes very widely spread, from the large phagocytes in the capillaries of the liver, spleen, and bone marrow. (The same cells which dispose of the aging red corpuscles.) If the agencies already committed to the defense are inadequate and the bacteria are winning in those scattered engagements, the body has still another line of defense—but not against every type of invading organism.

It has been part of man's knowledge for a long time that there are diseases from which a survivor rarely suffers again. Apparently that first attack leaves him with a receipt that he has paid his tax of pain to life and he is spared further visits from the same tax collector.

Induction of a mild case of smallpox in order to gain immunity to a severe attack was practiced a score of centuries ago by medicine men in India. They obtained pus from a patient with a mild case of the disease and smeared it into a scratch on a healthy person. The practice of such preventive inoculations against smallpox continued in the East but was introduced into Europe only in 1718 by Lady Mary Wortley Montagu, the wife of the British Ambassador in Constantinople. She had her son inoculated by a Turkish doctor. The wide prevalence of the disease spurred others to follow the example set by the courageous ambassadress and the practice of inoculation against smallpox became widespread. During the Middle Ages, to be free of pox marks was considered a mark of beauty. Little wonder that people were ready to subject themselves to the inoculations, even though the outcome was not always predictable. Sometimes the induced disease was a severe case of smallpox.

Chance taught us a safer and equally effective mode of protection against this disease.

Cows, too, are susceptible to smallpox. English farmers knew that cowpox was contagious among humans and considered the rather mild disease as an occupational hazard of dairymaids and others who came in close contact with cows. Who made the first observation that a case of the mild cowpox protects the sufferer from the virulent human smallpox is not certain. Some English farmers have been credited with this astute and profoundly useful correlation. But debates on priority, whether of ancient or of current discoveries, are unprofitable. The discovery is of far greater importance than the ego of the discoverer. Certainly it was Dr. Edward Jenner who established the facts with well-documented evidence and demonstrated how to harvest the benefits of the chance discovery. Since cowpox was known technically as *Variolae Vaccinae*, the purposeful inoculation with cowpox pus to immunize against smallpox came to be known as vaccination.

Vaccination became a very widely applied and gratifyingly effective measure of protection against smallpox. But the mechanism which produced the immunity of course remained unknown. In those pre-Pasteur days even the cause of infectious diseases was unknown. Pasteur demonstrated that some diseases and the immunity to them were induced by the same infectious agent. He did this in a celebrated public, scientific demonstration.

Anthrax was decimating the cattle herds of Europe. Pasteur traced the cause of the disease to the tiny rod-shaped bacilli which were teeming in the blood of the diseased farm animals. He also found that these bacilli could be rendered less virulent by keeping them, for a while, at temperatures much higher than normal body temperatures. (Most bacteria have become so accustomed to the warmth of the animals they inhabit that they live best at that temperature even though they themselves are unable to maintain such warmth.) The heat-treated germs would not kill the animals when injected. They merely made the animals ill. However, after the animals recovered from their mild attack of anthrax they were immune to the virulent organisms. They could shake off doses of anthrax bacilli which would surely kill unimmunized animals.

When the conservative physicians scoffed at Pasteur's claims he arranged a public demonstration which had all the trappings of a country fair. Scientists, physicians, dignitaries, and newspapermen all gathered in a field where fifty sheep, a few bottles of bacterial broth, and Pasteur were the center of attention. His assistants injected twenty-five sheep with a heat-weakened bacterial suspension. Twelve days later those same animals received a stronger dose of infection, consisting of bacilli which were exposed to less rigorous heat treatment. The sheep survived this injection, too. Finally all the animals, including the twenty-five "controls" which until then had been untouched, were injected with the same deadly dose of anthrax. Two days later, twenty-two of the controls were dead and the three others were in their final agony. Every one of the sheep protected with the heat-weakened bacilli was grazing contentedly!

Pasteur thus proved not only that infectious diseases are caused by microorganisms but also that those very germs rouse the animal to rally to its own defense and make weapons to repel future invasions. (Of course, today we know that all infectious diseases are not caused by microorganisms. Some, including smallpox, are caused by viruses. But only several years after Pasteur's death were viruses discovered.)

The nature of the weapons of immunity remained obscure for a long time after Pasteur's dramatic success with the attenuated anthrax bacilli.

The allergies which bedevil so many of us are also our saviors from germs. Outbreaks of hives, sneezing, the necessity for abstaining from certain foods, are relatively small prices to pay for having the weapons with which to combat infectious disease. For the same process which visits upon us the discomforts of allergies brings to us the blessings of immunity to disease.

Allergy means simply an "altered reaction." If we inject into a guinea pig some egg white from a hen's egg nothing unusual will happen. But if after some time we inject into the same guinea pig an identical dose of egg white there will be a violent reaction. The animal's breathing will become labored, it will thrash around, and will finally go into shock, from which—depending on the doses injected—it may or may not recover. The guinea pig has an altered reaction or an allergy to the second injection of egg white. What brings about this violently different response to the second injection?

Animals resist the entry of foreign proteins into their tissues. The proteins we eat do not normally enter our tissues intact. These proteins are broken down by the enzymes of the alimentary canal into their component amino acids and only these are permitted to enter our tissues. From the absorbed amino acids we fashion protein molecules in the image of our own proteins. (In patients who suffer from the various allergies there seems to be a minute amount of seepage of the foreign proteins into their tissues.) Therefore if a foreign protein does enter into the tissues of an animal it means that a stranger is within the gates. The stranger may be just a foreign protein molecule or a whole organism with its foreign proteins. The reaction of the animal is the same to either danger. It begins to fashion shock troops, or antibodies, in an attempt to dispose of the invaders.

Antibodies dispatch the invaders to their doom in a variety of ways. They can dissolve the cell walls of bacteria and the tiny monsters just ooze away; or they merely stimulate, by their very presence, the white cells to greater efficiency. Finally, antibodies combine with the intruding protein or germ, the so-called antigen, to form with it an insoluble particle. Once the foreign matter is thus clumped together the scavenging white cells and large phagocytes dissolve them at their leisure.

For example, in the blood of the guinea pig which has been injected with the egg white there appears an antibody which when added to a solution of fresh egg white curdles it. The violent symptoms of the guinea pig on the second injection are caused by the excessive curdling between egg white and antibody within the animal.

All antibodies are amazingly specific. The antibody from the blood of this guinea pig will not curdle the egg white from a duck egg or a goose egg as effectively as it does the hen's egg white. It is easy to see why such remarkable specificity of antibodies is essential. An animal would be in a dire predicament if it produced antibodies which curdled any protein at random. The antibody might curdle the animal's own proteins.

That antibodies are so specific is a great asset. We can take the antibodies formed against diphtheria by a sturdy horse and fortify a human child with those very antibodies. Or we can tell whether a brown stain on a cloth is ox blood or human blood. When the dissolved stain is mixed with the serum of a guinea pig which had previously received injections of human blood, the formation of a precipitate identifies the stain as human blood too.

Our practical knowledge of immunochemistry—that is the name given to the study of the antigen-antibody reaction—fills books. But our knowledge of *how* an antibody is formed is very limited. The problem simply stated is this: How can one protein molecule, the antigen, stimulate the shaping of another protein molecule, the antibody, so that united they form an insoluble curd, while each of them separately is very soluble? Before we can answer that we must first have answers to some of the most profound riddles in biology. How is the highly specific structure of a protein molecule encoded within a cell. How is that code translated into dynamic action so that identical new protein molecules can be shaped? Once we have answers to these questions we might then approach the problem of how a foreign protein molecule induces the cell to elaborate a brand new molecule of defense whose structure is so constructed that it can vitiate the foreign protein by embracing it in an insoluble complex. We shall take up what we know of protein synthesis in the next chapter.

We have seen earlier that the body is in a constant state of flux. Tissues are broken down and rebuilt constantly. The amino acids which compose our proteins today will be gone tomorrow and replaced by more recent amino-acid arrivals from our food. Are antibodies an exception to this constant building and dismantling? They are known to remain in the body for years after the infection which caused their formation has subsided. Indeed, sometimes they remain for a whole lifetime. Are antibodies permanent islands in the constantly changing sea of the body?

Dr. Michael Heidelberger, the foremost immunochemist, answered this question in collaboration with the late Dr. Rudolph Schoenheimer. They found that antibodies are no different from other proteins of the body. They, too, are being constantly assembled and dismantled. In other words, the machinery which is set into motion at the time of the original infection to produce antibodies, continues its production, sometimes for years, sometimes for a lifetime. We know then that the machinery keeps functioning. But what that machinery is, we have no idea.

Although I have emphasized the unsolved problems in immunology, actually the practical achievements in this field are enormous. The multitude of different immunizations, the classification of human blood into various types to insure safe blood transfusions, the diagnosis of diseases (such as syphilis) by testing a few drops of blood, the whole branch of medicine treating the allergies are the bounties harvested from the studies of the antigen-antibody relationship. But the discussion of those achievements can be left to others. To some of us the unknown is far more fascinating than the known.

Bacteriological warfare is a new term, much bandied about lately. It refers, of course, to the use of bacteria or their poisons as weapons in human warfare. But bacteria, too, have been carrying on warfare on a vast scale, millions of years before man and his puny wars became part of the earth's landscape. Man is a fumbling novice in the techniques of mass slaughter. What is it, a hundred thousand humans that we killed with one atomic blow? In less time than it takes to say "hydrogen bomb," millions of bacteria are the casualties in the warfare that goes on in a handful of garden soil. And the weapons used are at least as ingenious as man's.

In that handful of soil are more microorganisms than there are humans on the face of the earth. Life is hard in that handful of a world. Food is scarce; the struggle for survival is fierce. Some of the combatant microorganisms are especially well equipped in their battle against their competitors. They pour out a poisonous solution all around them. In the area staked out by the spreading molecules of the poison, no other organisms can enter and live. The wielder of the poisonous weapon can grow and multiply in his befouled homestead.

It is almost superfluous to say that it was Pasteur who first observed this Lilliputian chemical warfare. A batch of the anthrax bacilli which were described a few pages back stopped growing. He pinned the responsibility for the mass murder on some stray microorganisms which drifted into his cultures from the air.

In the garden of the mind of that genius, this chance observation became the seed of a dream. He visualized the slaughter of the pathogens which cause our diseases by the introduction of their natural enemies or their products into the patient. He wrote in 1877 that such a scheme "justifies the highest hopes for therapeutics."

It has taken sixty years for that dream to come true. There were little episodes which kept the dream alive during those years. In 1885 Cantani tried to cure a tuberculous woman by the inhalation of a nonpathogenic organism. He was apparently fairly successful. (It may have been a case of a normal recovery from tuberculosis.) At any rate the work was not followed up.

However, many bacteriologists continued to notice the lethal antagonism of some microorganisms to each other. The disappearance of pathogens from water which trickles through the soil became well known.

The destruction of one microbe by another came to be known as antibiosis. At the start of this century the first antibiotic— a substance extracted from one microorganism to kill another —was prepared. However, it was not successful.

The credit for the first isolation of an effective antibiotic goes to Dr. René J. Dubos, a bacteriologist, of the Rockefeller Institute and to his biochemist associate, Dr. R. D. Hotchkiss. Unfortunately their antibiotic, while very effective against bacteria, is also quite poisonous if injected into the patient. It is not used widely, except for surface applications, nor has it received the wide public recognition it deserves.

Dubos's work is the model in the search for all antibiotics. He took a pinch of soil and sprinkled it onto a glass dish coated with nutriment for bacteria. He separated the various strains of bacteria which grew and replanted them into little bacterial gardens of their own with lots of food. Having feasted them, Dubos put the bacteria to work. He placed a growing culture of pathogenic organisms—staphylococci, which cause boils—into each colony of soil bacteria. One strain of the soil bacteria refused to share its food with the staphylococci. These bacteria exude something which kills competing organisms.

The isolation of an antibiotic, later named gramicidin, followed the usual pattern of such searches. The poison-bearing bacteria were grown in large batches and the antibiotic was concentrated from their juice, using the increasing toxicity of the preparations to the staphylococci, as the guide in the various steps of the search. The task was completed in 1939. However, the antibiotic turned out to be quite poisonous when injected into animals. Nevertheless, it is a valuable aid in the treatment of exposed infections. But there were other, better antibiotics to come.

Streptomycin was dug out of bacteria by Dr. Dubos's former mentor, Dr. Selman Waksman of Rutgers. This is a rare case of the master following in the footsteps of his disciple.

Waksman has devoted his life to the study of soil microorganisms. They are of enormous importance agriculturally, economically, and even aesthetically. About a ton of leaves falls on each acre of forest every year. If it were not for the soil organisms which decompose all such debris, our earth would become a cluttered, uninhabitable graveyard in a very short time. No dead plants or animals would decompose. The substance of their bodies could never be returned to usefulness. Our accumulation of ancestors for thousands of years back would be with us perfectly preserved.

Waksman had been interested in soil organisms from the point of view of their value in agriculture. But after Dubos had extracted an antibiotic from such organisms, Waksman, too, channeled his efforts in a similar direction. Streptomycin is but one of the many antibiotics he and his associates extracted.

And now a few words about the greatest antibiotic of all those that have been isolated thus far—penicillin. In 1928 a mold spore drifted from the air onto a bacterial culture plate of Dr. Alexander Fleming, a bacteriologist at St. Mary's Hospital in London. The spore reproduced on the spread of food in the glass dish. As it grew, it exuded something, for there was a halo around the mold, an area free of the staphylococci which were the original inhabitants of that plate. Now, such accidents must have happened to scores of bacteriologists. But they would simply throw a ruined plate into an antiseptic bin.

Fleming had been carrying on bacteriological warfare—the proper kind, *against* bacteria—for a quarter of a century. He became interested in this mold which could parachute into a colony of pathogens and could slaughter them with ease. He transferred the mold to culture plates and grew it in large batches. The mold-free juices he prepared were still lethal to bacteria. Since the mold was a *Penicillium* Fleming named the antibacterial substance in the juice penicillin, after the parent. Fleming recognized immediately the potential value of penicillin. He wrote in 1929, that "It may be an efficient antiseptic for application to, or injection into areas infested with penicillin-sensitive microbes." But, as he wrote more recently, ". . . I failed to concentrate this substance from lack of sufficient chemical assistance. . . ."

Not until ten years later was penicillin concentrated and used in human therapy. Dr. H. W. Florey, a pathologist at Oxford, undertook in 1938 a systematic search for antibiotic substances. His biochemist associate was Dr. Ernst Chain, a refugee from Hitler. The team of Florey and Chain was spectacularly successful in winning penicillin. Their methods were communicated to American laboratories and pharmaceutical houses under the auspices of the Office of Scientific Research and Development. "The Americans," wrote Fleming, "improved methods of production so that on D day there was enough penicillin for every wounded man who needed it. . . ."

There is beautiful, fairy-tale justice in Chain's career. Exiled from his home, he became a key man in the fashioning of a drug which protected millions of young champions as they sallied forth to liberate his homeland.

Why was there a lag of ten years between the discovery and the perfection of one of the greatest drugs? Was it Fleming's fault? Certainly not. He was trying to enlist aid and interest; he sent filaments of the mold to any one who wished to grow it and study it. Was it the fault of scientists in general? No. As we shall see later, they can hardly be blamed. Who then is to be blamed that during those ten years men perished by the tens of thousands from infections which, with the aid of penicillin, they could have conquered? But let us leave this to the last chapter where the lag of ten years will be the leitmotif of a discussion on the neglect of research.

How does this most potent of cell defenses—which we borrow from molds for our own defense—kill bacteria?

The answer came slowly and piecemeal—as do all answers to questions in biology. First of all it was found that penicillin is lethal only to growing bacteria. If a bacterium is not growing, because of the lack of some nutrient, then it can survive exposure to penicillin. Dr. Hotchkiss, whom we met a few pages ago, made the first observation that bacterial cultures in presence of penicillin accumulate amino acid complexes. Then a young chemist, Dr. J. T. Park, working at the U. S. Army Bacteriological Warfare Research Center, made a penetrating observation. The amino acid complexes excreted by bacteria which Hotchkiss observed were attached to a nucleic acid fragment. Meanwhile other biochemists, mostly in England, whose interest was the structure of the cell walls of bacteria, found after painstaking analysis that the cell walls of bacteria are composed, in part, of the same complex which accumulates in the presence of penicillin. The mosaic now started to fit into a picture: penicillin causes cell wall components of bacteria to accumulate. The next large piece of information was contributed by a Swede, Dr. Claus Weibull. He was studying the mechanism by which an enzyme called lysozyme destroys bacteria. This enzyme which can dissolve bacteria was originally discovered as a component of tears by Dr. Fleming himself. This story has enough improbable coincidences to sound as if it were a plot spun by Dickens. But, actually, this is not unusual in the history of science, since those engaged in scientific pursuits have been so few. (Someone has estimated that 95 percent of all the scientists who ever lived are alive today.) In turn, those who have made lasting contributions are fewer still, and, therefore, several achievements by the same person in the same area are not unusual.

Dr. Weibull observed that in the presence of lysozyme bacteria just vanish in an ordinary culture suspension. However, if he supplemented the culture fluid with large amounts of salt or cane sugar the bacteria did not vanish but merely lost their shape; instead of their characteristic rod-shaped structures they appeared as shapeless globules. This finding was a clue on how to observe penicillin at work. The one who recognized the clue is Dr. Joshua Lederberg, one of the most brilliant of contemporary microbiologists, with whose major work we will become acquainted in the next chapter.

Dr. Lederberg allowed penicillin to attack bacteria in a culture medium into which he had incorporated cane sugar, after Dr. Weibull's example. He observed that in the presence of the sugar the bacteria did not dissolve away as they usually do but merely assumed altered shapes, they too became round globules. If these globules were removed from the environment of penicillin, rod-shaped structures slowly emerged from them. The mode of action of penicillin now became clear. A bacterium is but a minute speck of protoplasm awash in a vast, hostile sea. To guard the precious protoplasm against dissolution and to give them some structural strength bacteria have evolved cell walls. Penicillin interferes somehow with the production of those walls, that is why the cell wall building units accumulate in their culture fluids. Unaided by their walls the newly growing bacteria pop like bubbles in a churning sea. Sugar or salt can protect the bacteria by reducing, through a physical phenomenon, the pressure accumulating within the naked bacterial globule. It now became clear why only growing bacteria are killed by penicillin. A bacterium whose cell wall is fully formed is beyond the reach of penicillin which can kill only by preventing cell wall formation.

The reason for the complete innocuousness of penicillin for animals also became apparent. Since animal cells have no walls, penicillin can do them no harm. (The rare complications in penicillin therapy are due to allergic reactions.)

Penicillin is truly a wonder drug. But its wonder does not lie merely in its therapeutic effectiveness. Of equal wonder is the finding of a drug which pounces with awesome efficiency on possibly the only function different in bacterial and mammalian cells, the erection of a cell wall.

We know the chemical structure of penicillin and have even made it in the laboratory. How simple that sounds! That brief sentence summarizes years of work during the Second World War by scores of the best organic chemists of England and America. This Allied team was led by two of the ablest generals of organic chemistry of the two countries: Sir Robert Robinson, who came to this campaign after brilliant victories in the field of the structure of natural pigments, and H. T. Clarke, who decoded the structure and achieved the synthesis of the sulfur-containing part of vitamin B_1.

While the synthesis of penicillin is a brilliant achievement of this international team, so far it has no practical significance. It is cheaper by far to let the molds do their chemistry and make it for us.

We do not know the specific function of penicillin within the mold itself. We know altogether very little about the biochemistry of microorganisms. With excusable narcissism we have concentrated so far on the study of human biochemistry or of animal biochemistry related to it. Penicillin may be a normal metabolic product of the mold which happens to be poisonous to other microorganisms; or, it may have no role within the cell other than self-defense. A chance mutation, resulting in the ability to make penicillin, may have enabled that particular mold to survive, for, nature achieves her purpose in many apparently purposeless ways.

The parallel history of penicillin and of Dubos's antibiotic illustrates the role of chance in the rewards of research, too. Two groups of scientists, Fleming, Florey, and Chain on one hand, and Dubos and his team on the other, set out on a hunt for an antibiotic like so many prospectors hunting for gold. The prospectors have the same training, the same skills and tools. The team which succeeded first later found fool's gold mixed with the gold: the antibiotic was toxic. The other team, by chance, found pure gold, penicillin, and received accolades and Nobel Prizes. But, well has it been said that "The object of research is the advancement, not of the investigator, but of knowledge."

*Life's mystery is not truly manifested
in adult forms, but, according to my
way of thinking, resides essentially
in the reproductive cell and in its
capacity for future development.*

PASTEUR

10. GENES

THE BLUEPRINT OF OUR CELLS

"GLORY to the great friend and protagonist of science, our leader and teacher, Comrade Stalin!" With this fulsome dedication ends the book *The Science of Biology Today*, written by Trofim Lysenko, the winner in a gruesome struggle among Soviet biologists. The outcome of the struggle was not too difficult to foretell. Comrade Stalin was in Lysenko's corner, or, to be more accurate, Lysenko was in Stalin's corner.

What the defeated biologists think of the "great teacher" is difficult to report with accuracy; *their* books must await the time when the ultimate in Five Year Plans brings better publishing facilities to Siberia.

The crux of the controversy in which Lysenko emerged victorious is simply this: Can the characteristics acquired by a living thing in its lifetime be transmitted to its offspring?

It is at first surprising that an academic question which has been debated for about sixty-five years should become a political issue. Three factors raised the moot problem in biology to an affair of state: According to Marxist theory, a change in man's political environment could purge him of his base acquisitive instincts.[1] It is doubtful that Stalin would have been aware of such academic questions in biology if ambitious biologists had not brought it to his attention. And finally, the totalitarian state cannot tolerate the slightest deviation from the true path. It might become contagious.

The mechanism of heredity which has fascinated biologists for generations, has recently been partially revealed. Needless to say, it is a chemical mechanism. But, before we unveil this latest masterpiece shaped by the tools of biochemistry, let us review the groundwork of generations of biologists which forms the pedestal on which this achievement stands.

That the gross, species-characteristics of the parent are handed down to the offspring through the seed has been obvious to man for scores of centuries. The farmers of antiquity, planting their grains along the Nile, the Euphrates, and the Yangtze, confidently expected to harvest the same kinds of grains.

That the variations among individuals of the same species can be handed down to the offspring too, was also part of man's knowledge for a long time. The Arabian racehorse breeders understood the importance of heredity in improving the speed of their horses so well that they kept elaborate pedigree records for centuries. Indeed no race horse could be labeled a purebred Arab unless its ancestry could be traced to one of five famous progenitors of the Arab breed.

While it was recognized that certain traits are passed down from parent to offspring, the patterns—indeed the laws—which heredity follows remained unknown until the middle of the last century. Even then their discovery remained obscure and the announcement of the discovery itself had to be rediscovered thirty-five years later.

[1] The ennobling Soviet atmosphere has as yet made no dent in the self-promoting instincts of these biologists.

Gregor Johann Mendel chose for himself a life of peace and obscurity as an Augustinian monk. But for his hobby, he would have remained one of the anonymous brothers of the monastery at Brunn, Austria. However, he pursued his hobby, in the tiny monastery garden with such brilliance that he was lifted from his self-imposed anonymity and he is counted among those men of rare genius who are the first to discern a law of nature. Mendel was the founder of the science of genetics. He shaped the tools for the study of the ways of heredity.

He studied the effects of the crossbreeding of two plants endowed with contrasting traits. He started by planting different peas, tall and short, in the monastery garden. If these two varieties of peas were allowed to self-fertilize, with no possibility of cross-fertilization, the plants yielded seeds which bred true to type: they grew into plants, tall or short, like their parents. But if a tall and a short pea were cross-fertilized or "hybridized," the seeds from such a union gave rise only to tall peas. Mendel did not stop there. He patiently crossed these tall peas of mixed ancestry and collected *their* seeds. Out of these seeds grew both tall and short peas. He carefully recorded his seeds and crops and found from over a thousand different plantings that the tall and short "grandchildren" always appeared in a definite ratio: 75 percent tall and 25 percent short. He repeated these studies with red- and white-flowered peas. When he cross-fertilized these flowers he found that all the seeds gave rise to red flowers. But these second generation red flowers, when cross-fertilized, produced seeds from which grew both red and white peas. He found 6,022 red and 2,001 white flowers, again a ratio of three to one (75.1 to 24.9 percent, to be more exact).

Mendel concluded that there are factors in peas which determine their color and height. The factor for whiteness or shortness remains dormant in the first generation after the crossbreeding of opposing traits, but asserts itself in the second. Moreover, the dormant factor reappears in the second generation in a ratio of one to three of the dominant factor.

This profound discovery lay hidden for thirty-five years in the pages of the local scientific journal in which Mendel published it. The eyes of the whole scientific world were finally focused on the great discovery when in 1900 three different biologists simultaneously stumbled on the life work of the patient monk. Building on this foundation, other biologists quickly erected the science of the study of heredity, genetics.

The next great experimental approach was made by Dr. Thomas Hunt Morgan, who undertook to unravel the hereditary maze of an insect, the fruit fly. He found that the fruit fly has a great many traits inherited in a Mendelian way. Some of these traits tend to appear together and four such groups of linked traits could be identified. It was known that in the cells of the fruit fly there are four pairs of rodlike structures, chromosomes (colored bodies), whose function, however, was a mystery. Morgan surmised that these chromosomes might be the seat of the machinery which hands down Mendelian traits from generation to generation.

Since there are more traits than chromosomes and, since several traits are linked together, the conclusion was drawn that one chromosome houses the controlling mechanisms for more than one trait. These tiny, invisible fragments of the chromosome, each of which controls a single trait, were called genes. It is believed that a separate gene exists for every single trait.

The larger aspects of the mechanism of heredity now became clear. Every individual has two sets of chromosomes, one set contributed by each parent. We humans have twenty-three pairs; twenty-three from each parent. (The reproductive cells contain only one half as many chromosomes as the other cells, otherwise our chromosomes would be doubled every generation.)

To illustrate the mechanism of heredity let us return to the sweet pea. When a white and a red sweet pea are cross-fertilized, each sex cell contributes a gene for the control of color; one a red, the other a white. For some reason the red gene dominates over the white and such a hybrid seed will always give rise to red flowers even though it contains both the red gene and the white gene. In forming the sex cells of this hybrid plant only one of these genes, the red or the white, goes into each sex cell. Thus these sex cells contain the red or white gene in the same ratio, one to one. The probabilities of fertilization are as follows: A red male can fertilize a red female; the offspring from such a union will be red because both genes are red. A red male cell can fertilize a white female; the offspring from such a union will also be red because the red gene is dominant over the white. A white male can fertilize a red female, producing again a red offspring. Finally a white male can fertilize a white female producing the only white offspring of the four. This is the reason for the three to one ratio of red to white peas that Mendel recorded.

As far back as 1883 A. Weismann, a German biologist who was unaware of Mendel's work, differentiated between traits acquired by an individual during its lifetime and those which are handed down to it from the cells of its parents. Weismann thought that the acquired traits cannot be passed on to the offspring. To test his assumption he subjected experimental animals to drastic environmental changes. For example, he cut off the tails of mice for nineteen generations. He could observe no change whatever in the tails of the newborn mice from the mutilated ancestors.

It was known, however, that from time to time some novel traits which *are* passed down to the offspring appear in individuals. These freaks or sports appeared here and there by rare chance. To some of the inhabitants of a certain Swiss valley six-fingered babies are born; among the orange trees of Brazil a seedless, navel orange makes a sudden unexplained appearance; a silver fox gives birth to semi-albino offspring and launches the platinum fox industry. The biologists could but wait for these sports or mutations to occur; they could do nothing to induce them. They tried valiantly despite Weismann's failure. All kinds of environmental changes were devised to induce such inheritable mutations, but they all failed.

Finally, biologists resorted to stronger measures. Dr. H. J. Muller, one of T. H. Morgan's students at Columbia University, exposed the fruit fly to X-ray radiation. All kinds of freaks appeared among the progeny of these flies; extra wings, a lack of wings; extra legs, no legs. Biologists were delighted. The gene was approached at last. Apparently the intense energy of the X ray when it hits the gene target dislocates its structure, producing internal havoc and monstrous offspring.

Muller's man-induced mutations eventually enabled us to study how the tiny submicroscopic gene controls the shape, the color, the very essence of a living thing. A fortunate partnership between a geneticist and a biochemist was formed to explore this problem. Dr. G. W. Beadle and Dr. E. L. Tatum of California studied the mutations of a very simple organism, the bread mold *Neurospora crassa*.

This little mold makes very few demands on the world. It thrives if it gets sugar, water, some salts, and just a dash of one of the B vitamins, biotin. Out of these few starting materials the mold fashions an astonishing variety of substances for itself. It makes its own amino acids, its own vitamins, indeed anything else it needs for life. It owes its versatility to the large variety of enzymes it has for such tasks of synthesis. For every chemical process in living things is enzyme operated.

But if this self-sufficient mold is exposed to X-ray irradiation some of the next generation can no longer live on the simple diet of the parents. These offspring will die on a diet of sugar, salts, and biotin. Some will need one or more of the vitamins, others will need some amino acids. Only when these foods are provided will these crippled offspring live. And the dependence on these extra food rations is now handed down from generation to generation, proving that a mutation has taken place—a mutation which made the mold lose the know-how of making amino acids and vitamins. Since it was enzymes which made these foods in the natural *Neurospora*, we can conclude that these particular enzyme potencies must have disappeared under the impact of X rays. But these traits, which are transmissible from generation to generation, reside in the genes; therefore the function of genes becomes apparent: they control the fashioning of enzymes.

Before these brilliant discoveries were made we had hints that the genes control our enzymes. There is a rare type of idiocy among humans called oligophrenia phenylpyruvica. The unfortunates who are born with this affliction all have tiny skulls (oligophrenic); their intelligence quotient is anywhere from 20 to 50; and they are always very blond. These patients cannot metabolize completely one of the essential amino acids, phenylalanine. Since the body's pigments are made out of this very amino acid, the reason for the extreme blondness of these unfortunates becomes apparent. One of the intermediate products of the metabolism of this amino acid can be found in the blood and in the urine of these patients. Just as diabetics void sugar because they cannot handle it, these subjects excrete a product of amino acid metabolism with which they cannot cope.

The disease is definitely hereditary. Some ancestor of these patients lost the ability, through a ghastly mutation, to make a particular enzyme in the chain of enzymes which metabolize this amino acid. Since both synthetic and degradative processes are performed through a series of enzyme-motivated steps, if but one such a step is knocked out in the cell's assembly line, the synthesis—or the degradation—of a particular substance is stalled. Such enzyme-endowed abilities were gained and lost apparently at random, through the long evolutionary history of a present-day organism. There is no correlation between the position of an organism on the evolutionary scale and its synthetic ability. Some microorganisms, like the bread mold, have prodigious capabilities in fabricating amino acids and vitamins; others—for example, the lactic acid organisms—must obtain dozens of preformed essentials from their environment or they starve to death. Even among so homogeneous a class as the mammals there are wide discrepancies in synthesizing ability. The white rat cannot synthesize the amino acid histidine, but another member of that class, man, makes this amino acid with ease. There is recent evidence to indicate that the ability to fabricate a given compound might have been gained and lost several times during the evolutionary history of an organism. It is most probable that every synthetic ability was acquired by mutations. The chances are that the first ancestral living forms had no synthesizing abilities whatever. It is difficult to visualize how such primordial forms could have arisen *de novo*, equipped with the complete battery of enzymes needed to synthesize the compounds they required for life. The likelier assumption is that they could but forage, as it were, on the multitude of carbon compounds which must have been scattered in generous abun-

dance on the cooling crust of our once hot planet. The ability
to synthesize must have accumulated as the enzymes—or, rather,
the genes for the production of those enzymes—were slowly
acquired by fortuitous mutations. The source of the staggering
biochemical complexity of contemporary organisms now becomes clear. A microbe or a man is a biochemical potpourri,
a summation of those haphazard mutations.

There have been complaints from lay critics that the biological scientist is barren of imagination. He is accused of rarely,
if ever, making the sweeping generalizations of, for example,
the atomic physicist. The fault, if it is a fault, is not the biologist's, it is nature's. Either there is no over-all pattern to life, or,
if there is such a pattern, it is much too complex to be revealed
to our current, undeveloped minds. How can we predict whether
a particular enzyme is, or is not, present in a living thing which
is here today as the culmination of thousands of haphazard mutations? In man's blood and urine we find a substance, uric
acid, as a waste product of protein metabolism. In dogs, uric
acid is almost completely converted by an enzyme in their
livers to something else, allantoin, which they then excrete.
However, one breed of dogs, the Dalmatian coach dog, excretes
not allantoin but uric acid, as man does. We challenge the
greatest minds to tell us how we could have predicted a priori
that the Dalmatian coach dog would follow the human pattern
in respect to this waste product.

The knowledge of the biological scientist comes from cumulative experimental probings. He must patiently unfold the com-
plexities of the cell, he can rarely predict them. This explains
the dearth of infant prodigies among outstanding biological
scientists. The mastery of the techniques of his craft and the
accumulation of knowledge with those techniques takes so long
that the biologist is usually well into middle age before he

achieves eminence in his field. Not so the mathematicians and theoretical physicists, many of whom are hardly out of their teens when they make their niche-carving contributions. But their task is simple. They have staked out for themselves only the inorganic universe, the atoms, the planets, the stars. The biological scientist surveys the microscopic cell into which are packed all the forces and components of the inorganic universe *plus* the force and pattern of life.

How does the gene produce a protein molecule, an enzyme? Or, an equally vexing question, how does the gene duplicate its very self? We humans are launched into life from a single microscopic fertilized egg. This egg contains in its nucleus forty-six chromosomes which in turn contain an uncounted number of genes. Eventually every cell in our body contains forty-six identical chromosomes, presumably with identical genes. How did the genes duplicate themselves billions of times to populate billions of our cells?

What are genes made of? The answer came from an experiment completely unrelated to the question. Moreover, the interpretation of the finding was at first totally wrong. Scientific truth must often struggle to be recognized.

Dr. F. Griffith was a medical officer in the British Ministry of Health who, about thirty years ago, was studying the difference in virulence of various types of pneumococci. He isolated single bacterial colonies from the sputum of patients with lobar pneumonia, grew them separately, and studied their virulence by injecting the pure bacterial strains into mice.

He noted that the pneumococci which on special nutrient plates form little mounds of cells which have a rough surface were harmless when injected into mice, and those which form smooth, glistening mucoid mounds were virulent. To study the conditions under which a harmless culture may revert to virulence he killed a batch of smooth, virulent organisms with steam and injected this preparation along with live, rough, attenuated organisms into mice. The concoction proved to be lethal. When Griffith performed a bacteriological *post mortem* he came across an extraordinary finding. He found that the agents of death were live, smooth pneumococci which were teeming in the blood of the dead mice. In other words, the rough, harmless cells were transformed into smooth virulent ones. Griffith's explanation of his finding in 1928 may have been influenced by the preoccupation of scientists of that period with nutritional problems. He thought that the dead smooth pneumococci broke up in the body of the mouse and furnished a pabulum which the live rough ones utilized to build up a smooth structure.

But the molecular mechanism of this spectacular transformation in structure and virulence proved to be far more sophisticated than could be visualized by its discoverer.

However poorly Griffith interpreted his findings, he should be given full credit for describing a completely new phenomenon of profound potential significance. It is not easy to dig in the garden of biology where thousands of scientists have wielded their tools before and to unearth something truly new. Developing and refining phenomena previously recognized by others is relatively easy: Every one who has to eke out a career in science can and does manage just that. But to observe and lift out something really new requires special gifts. It requires a humility before nature and a hauteur toward one's colleagues. Nature's clues must be revered, prevailing dogma rejected. The scientist who can pioneer must have insight, he must know how to design a clear-cut experiment; he must have sufficient self-confidence to be sure that what he observes is real and sufficient mastery of the field to know that what he has found is new. And, finally, a dash of luck helps too.

Others who repeated Griffith's discovery found that the mice contributed nothing to the phenomenon. The same transformation from rough to smooth could take place by merely shaking live rough cells with killed smooth ones in a test tube with nutrient media. Moreover, even the killed cells were not needed. An extract prepared from them was also highly effective.

Finally, Griffith's observation came to the attention of one of the foremost microbiologists of the recent past, Dr. O. T. Avery of the Rockefeller Institute in New York. In a paper which is a classic, both for its content and style, Dr. Avery and his associates outlined their task in exploring Griffith's finding: "to isolate the active principle from crude bacterial extracts and to identify if possible its chemical nature. . . ."

Using the standard procedures for the extraction and isolation of a biologically active principle Dr. Avery and his associates obtained a substance from extracts of the smooth pneumococcus which was still able to transform rough to smooth cells at a level of about a hundred-millionth of a gram (10^{-8} g.).

Once the pneumococci were transformed to the smooth type by such a preparation they continued to divide and produce, indefinitely, smooth-coated organisms.

This was an extraordinary finding. A minute amount of a preparation from one type of cell could impose a hereditary change on the submissive recipient cells. What was this masterful substance?

The transforming factor—which is what Dr. Avery's extract came to be known—turned out to be a well-known component of all living cells: deoxyribonucleic acid, or DNA.

DNA has had a long history in biochemical research. It was

first isolated from the nuclei of pus cells in 1871. Since then it was found to be present in every type of living cell, concentrated in the nucleus when such a structure is present. Even its exact location in the nuclei had been pinpointed: DNA is a component of the chromosomes, the threads which seem to be associated with the transmission of hereditary characters from cell to cell.

Prior to the work of Avery the function of DNA was merely conjectured. We could not remove it from the cell and pin any function on it, a feat which can be done with scores of protein enzymes. But here for the first time was evidence for a role of DNA as the master molecule in whose structure may be locked such hereditary traits as the virulence and appearance of bacterial cells. Confirmation of this view soon came from the discovery of other transformable traits. For example, in any population of microorganisms some individual cells can resist the drug streptomycin. DNA extracted from such resistant cells can confer hereditary immunity to the drug to otherwise sensitive cells.

How does this master molecule dominate cells? A partial answer to this question has come as the harvest from some of the most ingenious biological experiments that have been performed in the past decade. Dr. Joshua Lederberg proved that very rarely, but definitely, sexual reproduction takes place in some bacteria. If we invade their privacy via a microscope we can observe in some strains that two bacteria, one with a male and one with a female tendency, can become paired; a minuscule tubule forms between them, and they remain attached this way for as long as an hour. That there is an exchange of genetic material between them was shown by Dr. Lederberg with an ingenious experiment which earned for him the Nobel Prize. Let us assume that there is a strain of male bacteria which cannot synthesize a vitamin and a female strain which cannot synthesize an amino acid. Neither of these strains can live without the appropriate dietary supplement. However, if we mix the two cultures and allow them to mate, they will have offspring which need neither of these foods. Such nutritionally independent organisms could have arisen only by the pooling of the synthesizing capacities of the two parents. In other words, once the know-how of amino-acid-making passes from the male into the female, which could make the vitamin all along, an organism in which both abilities are present may emerge.

Can DNA be compared to the conductor of a symphony orchestra who can command the various sections of the cell to perform at will; or is the chromosomal DNA thread more like an elongated blueprint in sections of which certain information is encoded? A clever variation of Lederberg's experiment performed by two French scientists, Doctors F. Jacob and E. Wollman of the Pasteur Institute, resolved this question. They chose bacteria for sexual mating with several nutritional, and therefore enzymic, deficiencies. They permitted the pairing off and the tubule formation to take place, but then at intervals they intruded, removed samples of the nuptial culture and whirled them in a Waring blendor. This barbarism severs the mating tubule but does not otherwise harm the organisms, they can continue growing and dividing. The experiment revealed that the longer the organisms were permitted to mate the larger was the variety of formerly missing enzymes which appeared in the offspring from the interrupted mating. In other words, the enzyme synthesizing potential does not jump from cell to cell at once but in a definite, timed sequence. The chromosomal DNA thus turns out to be a long tape on segments of which the blueprint for the shaping of each enzyme is encoded.

Recently a new type of transformation was achieved which confirmed the enzyme-shaping role of DNA.

If bacteria are exposed to ultraviolet irradiation, biochemical mutations, similar to those in the bread mold are produced. The bacteria lose the ability to synthesize some of their nutrients. They become biochemical cripples who must depend on their generous environment for an amino acid or a vitamin, otherwise they perish.

Such biochemical cripples can be restored to their original state of integrity by bathing them in DNA extracted from intact, non-mutated cells. In other words, a lost capacity is restored by the DNA from the complete cells.

How the DNA imposes itself on the cell and commands the shaping of the final product, the protein molecule with special enzyme function, is one of the unsolved riddles of biochemistry. We know that another kind of nucleic acid, ribose nucleic acid (RNA), is located mainly outside of the nucleus and, according to our current concept, it is this RNA which has the role of the building contractor which translates the blueprint into the finished product. (In some viruses, for example tobacco mosaic virus, RNA has the ambitious role of carrying the code and translating it into effect.) The molecular mechanism of the transfer of information from DNA to RNA and the mode of aligning of the amino acids by the RNA into a particular pattern are unknown at the present time.

What we do know is that the RNA does not act as a die which stamps out a protein molecule. We can detect the unassembled building blocks of the RNA molecule with different amino acids attached to them in the areas of the cell where the proteins are fashioned. This indicates to us that the RNA and the protein it shapes are assembled simultaneously and the protein then somehow peels away.

The speed with which a cell can align perhaps a thousand amino acids to shape a protein with a specific structure and a unique function is astonishing. It can be calculated that in a bacterial cell this takes a second or two. The wondrous efficiency of a living cell is a never-ending source of awe and admiration to those of us who are fortunate enough to spend our lives studying it.

We can no longer view the gene as a separate physical entity. It is merely a segment of the chromosomal DNA strand which is responsible for the shaping of a specific enzyme. The term, however, is still as useful as when it was first coined by the biologists. Its use avoids the clumsy circumlocution: The section of chromosomal DNA with a specific enzyme synthesizing potency.

The genes determine not only our structure but our very personality. No two human beings are alike. Differences in gross structures have been recognized for a long time: Identification by fingerprint takes advantage of just one area of individuality.

But recent evidence indicates that we are truly individualistic, even in more subtle ways. Dr. Roger J. Williams of the University of Texas, who, as we saw in Chapter 3 did brilliant work on vitamins forty years ago, has been exploring a fascinating new area recently: the biochemical basis of individuality. He and his associates measure in human subjects a variety of metabolic parameters. These include some metabolic components in the blood, excretion products in the urine, and individual reactions to certain drugs. From these measured entities they prepare a chart for each subject. This is in the shape of a star with lines of different length radiating from a central point. Each line represents a measured entity, the length of the line indicating the quantity. Thus a lot of information is concentrated in a small area and a sort of metabolic fingerprint is obtained for each subject. No two persons show the same pattern, indicating that we differ not only in gross structures but even at the level of our molecular components. The only exception to this rule are identical twins whose metabolic fingerprints are always essentially the same. Since identical twins originate from one cell and are thus genetically identical, and since it is only they who have similar metabolic patterns, we have evidence of genetic control at the molecular level.

Thus our genes determine not only whether we are tall or short, dark or light, but also how we metabolize our sugar or cholesterol, what is the level of hormones in our blood, how we react to nicotine—in short, they are the ultimate determinants of the many-faceted jewel that is life. Since this is so, the obvious question arises when will we be able to make that jewel shine brighter and longer by mastering our genes. Unfortunately, at the present time any thought of biochemical intervention with genes for therapeutic purposes is in the realm of science fiction. It took nature perhaps a billion years to so wondrously perfect the DNA molecule that strands of it can produce an Einstein and a Beethoven, and we have had only about ninety years to study it.

Since perhaps the very spark of life may be sheltered in the coil of the DNA molecule, its study is attracting investigators from several disciplines of science. Physicists, chemists, geneticists, all scramble hopefully in this area, and well they may, for the structure and mode of action of DNA is the greatest challenge to our ingenuity offered by any component of a living cell. The size of the DNA molecule is vast, running to six million times as heavy as a hydrogen atom. Thus, it is a thousand times larger than an active protein molecule such as insulin. The molecule's components are unique. There is a long core built up from phosphoric acid, from the core extend molecules of a sugar, desoxyribose, and to the sugar molecules are attached two different kinds of nitrogenous bases, purines and pyrimidines.

According to our current view the blueprint for a cell's machinery is encoded in the sequence of these nitrogeneous bases. Each time a cell divides the information contained in the DNA molecule must be duplicated to provide the two daughter cells with their master code. How this may be done has been explained by an ingenious hypothesis proposed by Dr. F. C. H. Crick of Cambridge University and Dr. J. Watson of Harvard. The catholicity of scientific disciplines which are focused on this problem is well illustrated by Dr. Crick's career. He is a physicist whose training for his current role seems to have been the design of aerial torpedoes during the Second World War.

Doctors Crick and Watson postulated from fragments of physical evidence that DNA exists as a double strand of two identical filaments which are held together by some weak interatomic forces. DNA might be visualized as a coiled zipper. During cell division the zipper separates, with one strand going to each daughter cell, there to be fitted with a new identical zipper mate.

That the making of the new DNA strand is entrusted to enzymes has been shown recently by Dr. Arthur Kornberg, a biochemist at Stanford University, who has received the Nobel Prize for his achievement.

He has been able to purify enzymes from bacterial cells which can induce the small molecular components of DNA to became tacked onto a DNA molecule which must be present as a primer. This is something like increasing the size of a snowball by rolling it in tacky snow. Of course, snow is snow regardless of its shape;

the precursors of DNA are only chemicals, but when they are woven into the pattern of DNA they become a structure of wondrous complexity and infinite potentiality.

This master molecule apparently shapes enzymes not only for the housekeeping tasks in the cell's functioning, but also for its own self-perpetuation. The control by the DNA of the shaping of just the right number and right kind of enzyme molecules for the making of just the right amount of DNA is a neat problem in cybernetics.

It is highly significant that Dr. Kornberg can find no production of DNA by his enzymes in the absence of traces of DNA.

There is another demonstration from a different source that the huge molecules of living things can be assembled only if there is a preexisting molecule to act as a dressmaker's pattern.

Starch is a huge molecule made in plants, such as the potato, by clipping together hundreds of smaller glucose molecules. Dr. Carl Cori and Dr. Gerty Cori, a husband and wife biochemical team, the third couple in history to be honored by a joint Nobel Prize, repeated the starch-making feat of the potato, but they made starch in a test tube. They purified an enzyme preparation which, they felt confident, should be able to join together the hundreds of glucose molecules into starch. All their attempts were fruitless. Finally they added just a trace of starch to their solution of enzyme and glucose and at once the glucose molecules began to be assembled into copious amounts of starch.

It is very probable that the fertilized egg, even the microscopic human egg, contains most of the protein and other molecules that the adult has. These molecules must be there as patterns for the shaping of the tremendous variety of substances which eventually make up the billions of cells of the adult. Since the molecules are infinitesimally small compared even to the microscopic egg, we can calculate that billions of such molecules can fit even into that tiny speck of protoplasm.

Twelve years have passed since Mr. Lysenko's harangue to his fellow biologists: "By ridding our science of Mendelism—Morganism—Weismannism—we will expel fortuities from biological science." During those twelve years great strides have been made in our understanding of the mechanism of genetics, but we have found no shred of evidence that an environmental stress outside of the DNA molecule can produce a hereditary change or that nature operates not on "fortuities" but on a Five-Year-Plan. Nevertheless, Mr. Lysenko marches on. He is apparently as powerful under Mr. Khrushchev as he was under Stalin. He dictates which Soviet scientists may attend international conferences and can handpick the editors of at least two scientific journals. And whoever can dominate editors, dominates science. Editors of scientific publications are the most powerful forces orienting research. Any discovery which no editor will publish might just as well not have been made. A scientist without a journal as his outlet is like a man who finds a treasure trove on a completely isolated desert island. An editor or any other power who smothers controversial or unpopular findings is ultimately smothering all of science.

Controversies in science are not unusual. Two groups of workers can very easily produce clashing fragments of evidence and interpretation as they grope in the intricate maze of a living cell. Nor are such controversies necessarily unhealthy. A discovery of a hidden secret of nature is a joyous event, but some scientists get an extra fillip if such a discovery at the same time demolishes those retarded colleagues who have the temerity to hold opposing views. Sometimes the pursuit of a

controversy to a triumphant conclusion is a zestful motive even
for the scientist who is completely consecrated to the pursuit
of truth. Such controversies are often spirited affairs. Sides are
taken by the onlookers; particularly deft thrusts by tongue or
test tube are cheered; the contestants are goaded. Only professional dignity is the deterrent to the placing of bets on the
outcome.

But when a supreme policy of a totalitarian state is en-
tangled in the controversy it becomes a grim affair. The state
cannot afford to lose.

How low the Russian authorities can stoop to bolster their
side is indicated by their conversion of an erratic Austrian
scientist, a Dr. Kammerer, into a hero of Soviet biology. Needless to say, he believed in the inheritance of acquired characteristics. Not only did he believe in it, he demonstrated it. He
claimed that he induced changes in the pigmentation of salamanders by changing the color of their environment. (Some
salamanders change their color, chameleon-like, enabling them
to blend into their background.) Kammerer claimed to have
produced permanent, hereditary color changes in his salamanders by keeping them against a background of the same color
for several generations. The ingenious geneticist enjoyed quite
a reputation until someone multiplied the minimum time
needed to raise a generation of salamanders by the number of
generations which Kammerer claimed to have studied. Unfortunately for his reputation, it was found that Kammerer's own
statement of the length of his studies fell far short of the time
required to raise all those generations. More damaging still was
the detection, by an American biologist, of India ink injected
under the skin of toads which had been induced by their "environment" to become black.

Kammerer fled, after his exposure, to the Soviet Union, where he committed suicide. According to the great geneticist Dr. Richard B. Goldschmidt there was a movie widely distributed in Russia in which the hero's exploits were patterned after Kammerer's career. The movie hero was framed by bourgeois villains who injected dyes into his experimental animals.

The bizarre aspect of the whole controversy is the tremendous fuss made about an academic argument. No free biologist would be the least bit ruffled if someone should come forth with evidence of a new Mendelian trait achieved by changes outside of the gene. Within the briefest possible time the claims would be tested in a dozen different laboratories and if they were confirmed, we would simply change our pattern of thought about the whole problem; we have changed our concepts before, when facts warranted it.

Indeed there are some lines of evidence indicating that in rare instances hereditary traits can be housed in the cytoplasm outside the nucleus. However, this so-called cytoplasmic inheritance has not been found to be subject to superficial environmental influence any more than are those inheritable traits which are shaped by the DNA of the chromosome.

We have not even begun to scratch the surface of the biochemistry of cytoplasmic inheritance nor, for that matter, the biochemistry of DNA. The relationship between the structure of DNA and the specific protein molecule it shapes is the most challenging problem before us.

One can only hope that the integrity of our DNA will be protected from the damage that may be raining on it long enough for us to complete our studies. For it is an asset and a drawback of DNA structure that it is sensitive, among other agents, to radiation. We owe our evolution from some primeval single-celled organism to the mutability of the DNA structure. Were the original molecules of DNA in such organisms un-

usually stable, such cells would invariably have given rise only
to identical images of themselves; evolution through mutation
and natural selection would have been impossible. The drawback to the mutability of DNA is that most mutations are
either lethal or deleterious. But nature, left to her own devices,
has managed a nice balance between the timing of the frequency of mutations and the selection of those with survival
value.

But of course such a view is a rather selfish solipsism. We
have evolved and survived. The many species which have become extinct because they were not endowed by mutation to
adapt to changes might have had a different view of nature's
efficiency in providing them with survival values.

We are not certain what causes natural mutations. Plants
and animals are exposed to all kinds of potent natural radiations. The radioactive minerals radium, uranium, and thorium
give off high-energy radiations as a flower gives off its scent.
(Since the radioactive elements keep disintegrating at a constant rate, we know that the quantities of such elements on
earth today are but a minute fraction of the amounts that must
have been here eons ago.) Cosmic rays are constantly bombarding us. Our earth is bathed in ultraviolet rays. A natural
mutation may very well be induced when a gene in a sex cell
of a plant or animal happens to get into the path of a packet
of energy from any of these sources. However, this is still in
the realm of speculation; the spontaneous mutations may be
induced by stresses, other than radiation energy, within the
chromosome. One current theory is that they are produced
chiefly by random collisions between molecules. Another possibility is that mutations are produced by the accumulation of
some chemicals as a result of metabolism.

The known mutagenic effects of ionizing radiations and the

increase of the latter on the surface of the earth as a result of our success at unlocking the core of the atom is a source of concern to many of us.

The history of science is filled with instances of our failure to assess the danger of a new discovery. Scores of people have lost their limbs and lives because of our failure to recognize the danger from X rays and from radioactivity. As recently as twenty years ago we were blind to the potential danger of the atomic accelerators. Several physicists paid for their lack of caution with the corneas of their eyes.

What the increasing amounts of radioactive material on the surface of our earth will do to our genetic material in the centuries ahead no one whose judgment is worth anything is willing to predict. According to one reliable estimate we have been producing the radioactive isotope of carbon (C^{14}) at a rate fifteen times faster than nature during the past four years.

The pernicious aspect of C^{14} is its longevity. If we were to stop all atomic detonations forever, still, even 10,000 years from now, the C^{14} content of the earth would be higher than it was before that literally earth-shaking day in 1945 when the first atomic bomb was detonated in the desert of New Mexico.

Since carbon forms the thread from which the tissues of every living form are spun, the excess C^{14} which we are producing by atomic detonations will inhabit our genetic material for millennia. Exactly what the effect of this excess radiation will be is a matter of some debate among scientists. Some feel that since it is only slightly above the background radiation which was in our bodies prior to 1945, its effect will be negligible. Others fear increasing incidence of leukemia, increased numbers of monsters among us, higher incidence of stillbirths, and there is a pervasive fear of menacing the very integrity of our genetic material. Whether such fears are justified or not may not be known for thousands of years. The thought that the Cassandras of today may be proven right a thousand years hence, when the damage may be irrevocable, is beyond horror. We may be perpetuating genocide on a scale which makes puny the recent monstrous deeds for which the word was coined.

However, there is very little doubt in the minds of many of us what may happen to life on this planet in case the disaster of a full-scale hydrogen bomb war should be visited on it. The survivors of the pyrolysis would sustain so much immediate and cumulative genetic damage from the immensely increased radiation that the human species might vanish forever. The requisite for survival will be a very stable genetic make-up and prolific capacity for reproduction. The horseshoe crab and the cockroach are the marine and terrestrial candidates for the creatures that may inherit the earth. The apt slogan for a hydrogen war may be "Make the World Safe for Cockroaches."

*Science moves but slowly, slowly, creeping on
from point to point.* TENNYSON

11. VIRUSES AND CANCER

SOME of the virus diseases and the complex of havoc called
cancer are the most monstrous scourges inflicted upon us.
Mounting evidence indicates that these two afflictions may
have more in common than horror: it appears that the viruses
may be one of the important agents which can divert normal
growth from its well ordered path into the chaotic jungle of
proliferation from which there is no return.

How did the idea that viruses may be the causative agents
in cancer originate? One of the most elusive goals a historian
faces is to trace an idea to its source. The facts of science are
easily tracked. As soon as they are well established—and sometimes sooner—they are exhibited in the learned journals for all
the world to behold. But the hesitant hypothesis, the intuitive
insight, which defies immediate testing, often passes down
through the corridors of time only as an echo from mind to
mind, and the source of the echo is all too often lost. The history
of the hypothesis that a virus may be the causative agent in
cancer is almost as old as our knowledge of the existence of
viruses.

What is a virus and how was it discovered?

Plants, as well as animals, are prone to infection by microorganisms, the only difference being that plants do not have
mechanisms for defense by immunity. A disease of the tobacco
leaf attracted the attention of the Russian bacteriologist Dmitri

Iwanowsky in 1890. He traveled to the Crimea, where the tobacco fields were infested with a pestilence which caused a mottling and withering of the broad green leaves. The Tatar tobacco growers of the region called it the marble disease. In the eyes of more worldly observers the mottling recalled a mosaic pattern, and that image gave the widely accepted name to this curse of tobacco growers: the tobacco mosaic disease.

Iwanowsky found that tobacco mosaic disease was infectious: the juice expressed from a sick plant when spread on a healthy leaf transferred the disease. This finding confirmed an earlier report by a Dutch bacteriologist who had also shown the disease to be infectious and who even claimed to have identified the infectious agent as a bacterium. His evidence for this was that the juice was no longer infectious after passage through two layers of filter paper. This turned out to be a technical blunder which doomed its reporter to anonymity.

Iwanowsky, who was apparently a skillful experimentalist, also tested for the bacterial nature of the infective agent by using an unglazed porcelain cup to trap the bacteria. He found that this device, which was known to act as an impenetrable sieve for all bacteria, permitted the infectious agent of tobacco mosaic disease to pass through it. He concluded that the infectious agent was some kind of dissolved toxin. He was bold but wrong in his conclusion. Actually, a soluble toxin which could infect and commandeer a healthy cell to produce more toxin was an even more revolutionary concept than the biological reality which was slowly unveiled during the next four decades.

Another bacteriologist, Beijernick, repeated Iwanowsky's experiment some seven years later and confirmed it. He called the infectious agent which could squeeze through the tiny pores of the porcelain a *contagium vivum fluidum*: a liquid infective agent. Soon the infectious agent of a disease of animals—the hoof and mouth disease of cattle—was shown to be similarly elusive. From then on the term "filterable virus" began to insinuate itself into our language,[1] for it was found that a whole host of diseases are caused by viruses. When we are invaded by the appropriate virus we may come down with rabies, smallpox, yellow fever, dengue fever, infantile paralysis, measles, mumps, influenza, virus pneumonia, or the common cold.

Some of the viruses are unusually stable; others which are more mutable have caused sudden flare-ups of devastating epidemics, such as the great influenza epidemic of 1918.

Occasionally what appears to be a new disease can be caused by a virus mutating to sudden virulence. Historically one of the most interesting of these diseases is the "sweating sickness" which swept through England, apparently for the first time in 1485, after the Battle of Bosworth Field. The Earl of Richmond, to strengthen his hand in his bid for the throne of England, imported mercenaries from France. According to Shakespeare, poor Richard III became grounded during the mélée and, even though he offered his kingdom for a horse—certainly the most spectacular trade-in deal in history for new locomotion—he left his head and crown on Bosworth Field. After dispatching the "bloudy Dogge" the victorious earl became Henry VII and marched on London, bringing his mercenaries and, with them, probably, the "sweating sickness." A fatal epidemic swirled through London like fire through a dry pine wood. The afflicted ran a high fever, became red in the face, and, as the name implies, sweated copiously. The disease often terminated in death within a couple of days. Three successive Lord Mayors of London succumbed within a brief time. The disease had

---

[1] Since filtration means the separation of a solid from a liquid by a barrier, "unfilterable virus" would have been the more accurate description of the minuscule pests.

been unknown until then, and after sporadic outbursts it vanished again in about eighty years.

Sir F. M. Burnet of the University of Melbourne, one of the world's foremost authorities on viruses, says in his book *Viruses and Man* that the sweating sickness was "almost certainly a virus disease." The late Dr. Hans Zinsser, who was an authority on typhus fever, said in *his* book *Rats, Lice and History* that the sweating sickness was "most likely typhus." And typhus is not a viral but a rickettsial infection. The opinion of experts on the diagnosis of a malady over an interval of five centuries tends to be no more congruent than on a current one.

How do we know that a virus has the capacity for mutation? Unequivocal proof for this came not from the study of viral diseases of plants or animals but, rather, from a completely unexpected area, the viral diseases of bacteria. For bacteria too have their own parasites preying upon them.

As Jonathan Swift wrote:

> So, naturalists observe, a flea
> Hath smaller fleas that on him prey;
> And these have smaller fleas to bite 'em,
> And so proceed *ad infinitum.*

(Little did Swift realize how close to the infinitely small can some "fleas" come: it would take about thirty million bacterial viruses to cover the dot over the letter i on this page.)

As often happens in the history of science, the bacterial viruses were discovered more than once. Their first discoverer was an English bacteriologist, F. W. Twort, who in 1915 observed cleared areas on agar plates which were otherwise densely populated with bacteria. He found that he could prepare filtrates from the clear areas and successfully reinfect fresh bacterial cultures with them. He offered as an explanation of his finding the possibility that he was dealing with a filterable virus analo-

gous to the viruses of animals or plants. Unfortunately Twort
did no more with his finding; the Great War then raging
pressed more urgent tasks on him.

The second discoverer of bacterial viruses—the one whose
name is associated with the discovery—was a Canadian bacteriologist working at the Pasteur Institute in Paris, Felix
D'Herelle. The impetus for the discovery was rooted in the
traditions of the Institute and indeed in the very personality of
that remarkable genius who founded it. Pasteur visualized a
research institute not as an ivory tower isolated from practical
problems but rather as a symbiotic structure where basic research could be nourished by the profit from applications of
knowledge previously won.

There is no better example of Pasteur's enterprise than his
attempt to control the rabbit menace in Australia. In 1887 the
government of the colony of New South Wales offered a reward of £25,000 for eliminating the rabbit pests which were
threatening to crowd out the human population of Australia.
Pasteur was attracted by the challenge and sent an assistant,
Loir, with a culture of chicken cholera bacillus, hoping to start
an epizootic among the rabbits. Unfortunately the venture
ended in a fiasco: Loir was not permitted to land with the culture. (Only sixty years later was Pasteur's plan of biological
control of the rabbit infestation successfully carried out in
Australia with the virus of myxomatosis.)

D'Herelle, too, was engaged in a practical problem, which,
however, led to the observation of a new phenomenon, indeed
to an unknown type of living creature. In 1911 he was studying
a bacterial infection of locusts in Mexico. To control the locust
pest it was planned to induce artificially among them this epizootic which was characterized by intestinal disturbances. To
do this effectively the bacillus which was the causative agent

was studied at the Pasteur Institute. D'Herelle discovered that sometimes the bacterial colonies had frazzled edges or, occasionally, clear areas, apparently free of bacteria, appeared in the midst of the otherwise continuous fields.

To account for this anomaly of growth D'Herelle proposed the hypothesis that the bacillus which he saw was not the true pathogenic agent but merely an associated organism. The real pathogenic agent, he conjectured, did not produce visible bacterial colonies. Its presence was made known only by the irregularities it produced in the pattern of growth of the associated bacillus.

D'Herelle started a long series of studies to see whether he could observe such dualism of infectious agents in human diseases. He focused on bacillary dysentery and typhoid fever. He tried to produce these diseases in animals by injecting them simultaneously with cultures of the pathogens and with extracts of the stools of humans sick with the same disease. Such extracts, he thought, might be a source of the associated organisms. At the same time he also looked for any effects the extracts from the stools might have on the growth of the bacteria themselves. And that is how it came to pass that a false hypothesis led to a real discovery. In August of 1916 a patient was recovering from a severe case of bacillary dysentery at the hospital of the Pasteur Institute. An extract of his stool was filtered free of any bacteria and this filtrate was added to a heavy growth of the dysentery bacillus. Next day the bacilli vanished! More bacilli were added. The newly added microorganisms were also disintegrated in about ten hours. Now only a drop of this clear fluid was added to a heavy culture of bacilli. Again a clear solution, as if there had been no bacilli present, resulted. With each transfer the killing agent became more potent as if it were feeding and fattening on its prey: hence the name bacteriophage, or bacteria eater. The bacteriophage behaved as a virus on filtration, it passed through the finest of filters. This suggested to D'Herelle that he was dealing with some ultravirus which destroys the host cells at whose expense it multiplies.

The phenomenon and its interpretation was welcomed in bacteriological circles with the accolade worthy of any truly novel discovery: it was promptly rejected.

It is not so surprising as it may at first appear that scientists should be reluctant to accept new ideas. A truly new idea is one of the most unpalatable impositions a man can inflict on his fellow men. A new idea assaults the vanity of the recipient. If it is a valid and worthy idea, why did he not think of it first? Obviously, therefore, every novel idea must be subjected to critical scrutiny. Moreover, the Pooh Bahs of science often wield great influence in determining the welcome accorded to novel ideas, and their mood is not always receptive. Often they achieve positions of eminence not on the strength of their own originality but rather by developing and enlarging the ideas of the generation immediately preceding. It is natural for them not to be too encouraging to upstarts who start hacking away at the ideological pedestals on which they are perched.

But it is just as well that new ideas must struggle for survival. Only in this way can the few clear insights be sorted out from the murky false ones. And if an idea is sound, it survives. A fruitful idea vis-à-vis its opponents is like sprouting grass against a granite block; ideas and grass prevail.

For about two decades after their discovery, much of the work on bacteriophages was concentrated on their possible role as therapeutic agents in disease. The plan of attack was simple: phages were sought and often found which attacked and destroyed pathogenic microorganisms in test tube cultures. However, when therapy with phages against the same pathogens would be attempted in animals, the results were disappointing. The possible exceptions were diseases localized to the alimentary canal, such as cholera.

The history of the phage research of this era repeats the history of most pragmatically oriented investigations. It is easy to set a pragmatic goal, but the path to it is tortuous and often completely unpredictable. Actually, infectious disease was conquered not as had been planned, by bacteriophage, but by chemicals extracted from other microorganisms and from the chemist's flask.

Until the fourth decade of this century work in the area of animal and plant viruses also centered mainly on effective methods of prevention and therapy of the diseases they produce.

But, in 1935, one man with one discovery changed completely the direction of research in this whole area. Dr. Wendell M. Stanley was a young organic chemist working at the Rockefeller Institute. The director, Dr. Simon Flexner, was one of those rare scientific administrators who, secure in his own ability, could surround himself with others of equal or greater ability and who saw his task as one of stimulation not of domination. At Flexner's suggestion Dr. Stanley started to study viruses. Only a few years earlier Sumner had successfully crystallized the enzyme urease, which turned out to be a protein. Stanley extracted tobacco leaves diseased with the mosaic virus and subjected the juice to manipulations similar to those which yielded the crystalline enzyme in Sumner's hands. His was an astute decision which yielded a stunning prize. Stanley obtained a crystalline material which appeared to him to be a protein, no different from say, egg white. According to all criteria this was a pure, homogeneous, noncellular substance which could rest harmlessly, apparently indefinitely, in a bottle. However, if a water solution of it, even at a dilution of one part per billion, was painted on a living tobacco leaf the mosaic disease was produced. What was this masterful substance? Stanley thought it was a simple protein. In this he was wrong, for his crystals were shown to contain large amounts of phosphorous —a tell-tale sign of the presence of nucleic acids. The virus turned out to be a nucleoprotein.

The discovery, in the words of the English biochemist, Sir Charles Harington, "administered an intellectual shock to science." Here was a noncellular, lifeless substance, essentially just a large molecule of nucleoprotein, which is capable of forcing a healthy plant cell to make not its normal components but this foreign incubus.

The discovery changed the course of biological thought and research. Perhaps its most important influence was that it brought the problem of the viruses within the intellectual horizon of physicists, chemists, and biologists, who took up the challenge, motivated not by an interest in clinical applications of virus research but by the sweet and compelling lure of a new frontier. Like true frontiersmen, these were a bold, imaginative, and unconventional lot. The rewards of their efforts have been truly impressive; new knowledge about viruses is pouring out in cataracts from the laboratories of the diverse disciplines.

Perhaps the largest increment in knowledge came from the bacteriophages. In 1940, with the aid of the electron microscope, we were able to see at long last these elusive little enemies of the bacteria. At a thirty-thousandfold magnification a bacteriophage alongside a bacterium is like a tadpole alongside a walrus. Photographs taken in sequence reveal the devastating effectiveness of the bacteriophage. Within a couple of minutes of the mixing of phage and bacterium the tadpole becomes attached by its tail to its large victim. About twenty minutes later the walrus is disintegrated as if it had been hit by a bomb, and a hundred fresh tadpoles dominate the debris-laden landscape. The ease with which a few bacteriophages can destroy whole flasks full of bacteria so rapidly now became clear. Bacteria divide and therefore double every twenty minutes in the progression 1—2—4—8. Bacteriophages, on the other hand, can, under ideal conditions, proliferate a hundredfold during similar intervals as follows: 1—100—10,000—1,000,000.

The experience gained from the studies of the tobacco mosaic virus was applied to the bacteriophages. They too were concentrated and isolated in pure form. They turned out to be a minuscule bag of protein filled with nucleic acid.

What happens when a phage makes a leech-like attachment to a bacterium has been ingeniously spied out by Dr. A. D. Hershey, a biologist at the Cold Spring Harbor Laboratory on Long Island. Phosphorus is an obligatory component of nucleic acids; sulfur is present in the proteins in the form of sulfur containing amino acids. Hershey grew two batches of phages, one with radioactive phosphorus and another with radioactive sulfur. This was done by incorporating radioactive phosphorus or radioactive sulfur into the diet of bacteria which were then infected with phages. The offspring of phages which emerged from the slaughter of the bacteria were thus labeled with the appropriate radioactive element. Dr. Hershey then infected separate batches of fresh bacteria with the differently labeled phages. He could now follow with a Geiger counter the path of the isotopes and the course of infection of the bacteria by the phage. Only the phosphorus got into the bacteria, not the sulfur. Therefore, only nucleic acid penetrates, not the protein. Thus the ability to capture the bacterial cell's machinery and to order it to make not its normal components but the substance and structure of the attacking phage must be invested somewhere in the nucleic acid. Once again, as in the case of the transforming principle, the master molecule proved to be nucleic acid.

Confirmation of Dr. Hershey's conclusion soon came from work with the tobacco mosaic virus. The protein and nucleic acid of this virus was chemically separated and each component was tested for its ability to infect healthy tobacco leaves. The nucleic acid alone was the infectious agent.

How does the virus nucleic acid subvert the host cell it invades? An unequivocal answer to this question was provided by Dr. Seymour S. Cohen, a biochemist at the University of Pennsylvania. Dr. Cohen discovered in the nucleic acid of some phages a unique component which is not present anywhere else in the living world. Not only is this species of molecule itself lacking, but even the enzyme for shaping it is absent from bacteria. However, very soon after invasion by the phage a new enzyme capable of performing this very task begins to accumulate in the subjugated bacterial cell. The phages, thus, not only "eat" bacteria, but they can order what is to be cooked.

We can but wonder at the ease with which a virus can enslave the much larger host cell it invades. But it must be recalled that the virus is uniquely equipped for the task. It is made of the same substance the genes are made of, and thus the intrusion of a virus is tantamount to the insertion of new genes. The virus enslaves the cell by capturing the seat of its government. The viruses are akin to genes not only in their ability to shape enzymes, but also in their ability to mutate. While all of the viruses apparently have the ability to undergo mutational changes, this is most easily seen, oddly enough, in the smallest of viruses, the invisible phages. The reason for this paradox lies in the characteristic pattern of destruction of bacteria caused by phages. If a large number of bacteria are seeded on a semisolid nutrient milieu, a thick "lawn" of bacteria will overgrow the surface. If, however, a few phages were also mixed with the bacteria, then small clear holes or plaques—D'Herelle called them *tâches vierges*—will be visible in the dense opaque growth of the bacteria. The plaque is usually circular in shape for the following reason. One phage infects one bacterium, and one hundred new phages emerge. According to the laws of probability, these are bound to emerge in a radial pattern. These hundred phages infect one hundred´adjacent bacteria, producing, theoretically, ten thousand phages. These, in turn, can infect ten thousand bacteria, and so the seeds of bacterial destruction spread radially until enough bacteria are killed to produce a round clear hole. All plaques do not have the same shapes or sizes however. Some are tiny; some are larger; some have well defined edges, as if they had been punched out by a drill press; others have serrated edges like a buzz saw; some are clear; some have a haze over them. If the phages from the different plaques are removed by merely poking the hole with a platinum wire, and are transferred to new bacterial plates, each type of plaque is infallibly reproduced. Therefore, the characteristic appearance of the plaque is a hereditary attribute of the phage. This is a unique case of the inheritance not of a particular shape or structure of an organism but, rather, of the pattern of the devastation it produces. This attribute of the phage is mutable: it can be changed by the same agents which cause mutation in living organisms.

Are the viruses living? The answer to this question is a semantic game anybody can play. If life is defined not in classical morphological, but rather in dynamic terms, as a system capable of producing order out of disorder at the expense of energy and possessing mutability, then viruses are alive.

Aristotle wrote over two thousand years ago: "Nature makes

so gradual a transition from the inanimate to the animate kingdom that the boundary lines which separate them are indistinct and doubtful." Between the fish and the land animals
stands the lung fish. Between the true mammals and the non-
mammalian forms is the Australian duck-billed platypus, whose
milk pours not from mammary glands but from sweat glands
and drips along its hair. The apes separate us from the rest of
the animals in the spectrum of life. We might consider the
viruses as bridging the gap between the nonliving crystals and
the complex living forms. Indeed, even the transition between
viruses and cellular microorganisms is not abrupt; the gap is
bridged by the rickettsia which have some of the structures and
a few of the enzymes of the microorganisms, but which, like
viruses, can live only as parasites, in contact with a living cell.

We do not believe that the present-day viruses are remnants
of ancestral forms from which other, more complex, living things
had arisen. Viruses are unable to reproduce except in intimate
contact with living cells. We have not been able to achieve
growth of a virus in any cell-free medium. They can but siphon
the energy and substance of their living prey; and they are highly
specific parasites: they can be reproduced only within the cells
of specific hosts. If the viruses are remnants of primordial, ancestral forms, how could they have survived during the eons
before their current hosts, of relatively recent evolutionary vin-
tage, had made their appearance? Viruses are thus, more likely,
degenerated remnants of previously more complex forms. Indeed they may be degenerated forms of some common ancestor
from which both the virus and the current host may have originated. Another possibility which has been suggested is that
viruses are renegade genes which break loose from the genetic
structure of a living cell and assume a marauding parasitic existence of their own. This "naked gene" theory of viruses, which

would explain both the origin and the specificity of viruses, was put forth in the early 1920s simultaneously by three unusually gifted men: H. J. Muller, who later discovered that X rays can induce mutations; B. M. Duggar, who in the eighth decade of his life became an outstanding expert in antibiotics and gave us aureomycin; and Eugene Wollman, who unfortunately was not permitted to complete his life's work.

It was Wollman who contributed most toward the probing of his hypothesis. Soon after the discovery of phages two opposing views about their nature developed. D'Herelle held that a phage is an invading parasite, that it penetrates its victim and proliferates, and that the progeny emerge from the crumbling hulk of the host. Another view, whose chief proponent was J. Bordet, also of the Pasteur Institute, was that the phage originated within the bacterium without external infection. As it happened, sufficient conflicting evidence turned up to support each of the opposing views. Some strains of bacteria were discovered which always had phages associated with them. Such organisms seemed to carry the seeds of their own destruction for, without the addition of any external infection, they were destroyed from time to time with the concomitant production of phages. Such bacteria were called lysogenic or self-dissolving.

Wollman postulated that in such bacteria some genes went berserk, broke loose, and started an independent existence as phages. He also thought that genes might be transmitted from cell to cell through the liquid habitat of the bacteria. This was a prophetic view which was confirmed about twenty years later with the discovery of the transforming principle.

Although Wollman's work on bacteriophages was not completed, he made a deep imprint on the field. Science can be mastered from books alone only at the most elementary level.

For the journey to the frontiers of knowledge an experienced, confident, and willing master is needed as a guide. A neophyte must not only acquire new skills, but he also needs help to sort out the fresh from the stale among his ideas, and among his findings the promising from the trivial. Wollman served as the link between D'Herelle and the current generation of scientists who are engrossed in phages, several of whom were trained in his laboratory. His work and his life were rudely terminated in 1943 when he and his wife and collaborator, Elizabeth Wollman, were seized by the Nazis in Paris and shipped to an extermination camp. To put a relative value on human lives would be to sink to the level of the creatures who made murder a way of life. But from the point of view of the development of phage research it would be difficult to think of a more grievous loss to the field.

Fortunately the work on lysogenic bacteria was taken up after the war by an outstanding microbiologist, a personal friend of the Wollmans, Dr. André Lwoff of the Pasteur Institute. Dr. Lwoff came to this task after a varied and fruitful career of investigations in microbiology. He was one of the pioneers studying the nutritional requirements of bacteria and had adumbrated from his findings the single-gene, single-enzyme principle. Dr. Lwoff decided to resolve first of all the moot problem whether there is a real association in a lysogenic organism between the host and the latent parasite, or whether the infectious agent happens to be merely a persistent fellow-traveler in such bacterial cultures.

It is technically possible to seize under a microscope a single bacterium and to place it in a fresh nutrient medium for culturing. Thus, a pure homogeneous colony of any microorganism may be started from a single cell. Lwoff selected at random a single cell of a lysogenic bacterium. He placed it in a fresh medium and waited for growth and subsequent cell division. He seized but one of the two offspring and replanted it. With virtuoso skill he repeated this nineteen times. In other words, he had an organism of whose ancestry he was certain for nineteen consecutive generations. Furthermore, it can be calculated that if his inoculum was always one-tenth the volume of each fresh culture broth, the original culture broth which might have contained the associated virus had become diluted ten-billion-billionfold, and thus the original virus must have been lost. Nor could the original virus have reproduced. Viruses can reproduce only within a living cell, and they always destroy their host as they reproduce. But in this case every ancestral cell for nineteen generations had been accounted for. Yet the bacteria from a fresh colony raised from such a pedigreed cell were still lysogenic; if deprived of oxygen or if permitted to stay in a culture too long they would dissolve and liberate virus particles. Furthermore, Lwoff found that he could induce virus production at will in lysogenic organisms. He discovered that if a fresh culture of such organisms is exposed to ultraviolet irradiation, within a brief time scores of virus particles will develop in every single cell.

This was an extraordinary discovery which was followed rapidly by another equally exciting one: chemicals which were known to cause mutations or cancer in animals also initiated virus production in lysogenic bacteria. An agent which causes cancer in animals can activate a latent virus to virulence in bacteria! This finding is laden with hidden meaning. When, in the next few decades, we shall learn just one-tenth of what the lysogenic bacteria are trying to tell us, many of us feel we shall be the masters of virus diseases and probably of cancer.

To be sure, microorganisms are not afflicted by cancer; thus the hope of extrapolating knowledge gained from lysogenic bacteria to cancer may appear to be the naïve dream of a sanguine amateur. However, biochemists and other scientists who can study life's processes at the molecular level are ever mindful of the teachings of comparative biochemistry. The pivotal molecular mechanisms in every kind of living cell, whether in a worm or a whale, are the same. Cancer of animals—and of plants—is a wild autonomous growth which, once it starts, defies the restraining processes of the host, which normally keep growth within the bounds of the organism's shape and size. Once a benign lysogenic virus is activated by irradiation or by cancer-producing drugs the proliferation of the virus runs as relentless and as autonomous a course as cancer. The identity of the inciting agents in the two different diseases and the similarity in the course they run are suggestive of some common basic mechanism. The light of knowledge about the lysogenic phenomenon in bacteria thus served to spotlight a long neglected corner of cancer research.

The hypothesis that viruses may be the causative agents of cancer has a history almost as old as our knowledge of the existence of viruses. Apparently the first one to suggest such a hypothesis was A. Borrel, a scientist working at the Pasteur Institute in the early years of this century. His was a hypothesis of desperation: He was really searching for a microorganism as the etiological agent in cancer and, having failed to isolate any microbe, blamed his failure on an elusive virus. This was a natural alibi for anyone working at the Pasteur Institute. Only twenty years earlier, Pasteur himself failed to isolate the germ which causes rabies for the simple reason that the etiological agent of *that* disease is a virus. Thus Borrel's hypothesis was nothing more than a cavalier conjecture. Novel, arresting ideas which are beyond the reach of experimental probings can flow with ease from any undisciplined, imaginative mind. New ideas which can be exposed to the critical scrutiny of a controlled experiment are rarer. And, rarest of all are those original ideas which survive the experimental traps set by those most resourceful of devil's advocates—a scientist's colleagues competing in his own field. That a virus may be the causative agent in tumor formation has occurred independently to other investigators besides Borrel. One of these had the skill, the perseverance, and the luck to bolster his hypothesis.

Dr. Peyton F. Rous was a young instructor of pathology at the University of Michigan in 1908. His work came to the attention of that astute talent scout, Dr. Simon Flexner of the Rockefeller Institute, who was himself a pathologist. Dr. Flexner invited the young man to join his group at the Institute, and his judgment was soon confirmed, for in 1911 Rous published a classic finding which eventually gave the whole area of tumor research a new orientation. Dr. Rous chose to work with a tumor which infests the breast muscle of the Plymouth Rock chicken. To test his hypothesis of viral etiology of tumors he excised such a sarcoma, ground it up with sand, and passed it through a bacterial filter whose efficiency he tested by proving that it held back known bacteria. He took these elaborate precautions to be sure that his filtrate was free of intact cells, for the transfer of tumors from animal to animal by the implantation of whole tumor cells is a relatively easy task. When Dr. Rous was satisfied that the filtrate from his tissue hash was free of whole cells he injected it into ten healthy chickens. Four of them developed sarcomas at the sites of the injection. The suggestion by Dr. Rous that it was a virus which transmitted the sarcoma was accorded by the contemporary pathologists the proverbial welcome of a skunk at a tea party.

The gloomy view of the time on progress against cancer was epitomized by the dictum of Dr. James Ewing, one of the foremost pathologists of that day: "Cancer will be understood and controlled only when life is understood and controlled."

Skepticism against Dr. Rous's discovery and hypothesis took many forms. The director of the Imperial Cancer Research Fund doubted that Rous was dealing with a tumor. His argument went something like this: "It cannot be a tumor since a causative agent was found for it."

Dr. Flexner, who loyally supported Rous, recommended that an article be written for a *Festschrift* [2] volume which was being prepared to honor the sixtieth birthday of Paul Ehrlich. Rous prepared the article and suffered the pain and ignominy of getting it rejected by the editors as unworthy of publication.

Dr. Rous struggled on in the wilderness of professional rejection. The techniques available at the time were inadequate for the isolation of a virus. He tried to prove the infectious nature of other tumors. He spent two frustrating years working with a tumor which infests the mammary glands of mice. In retrospect the reason for his failure in this instance is obvious. The demonstration of the viral etiology of the mouse mammary tumor was finally done with purebred mice of one specific strain.

Fortunately, in Dr. Rous's case longevity triumphed over professional hostility. In 1938 he witnessed the isolation of the causative agent of the Rous sarcoma. It is a virus which can infect and cause a tumor at the level of a trillionth of a gram (10^{-12} g.). (This corresponds to about 20,000 virus particles.)

Dr. Rous also lived to see the demonstration that scores of other tumors in plants as well as in animals are caused by viruses, and, more recently, he is witnessing a resurgence of the

[2] The compiling of a *Festschrift* or jubilee book is a tradition started by German scholars to honor an outstanding colleague at some chronological milestone of his life. Usually a book or a pamphlet is assembled from contributions by former students, associates, and colleagues, in which the contributors attempt to prove how much better they are than the honored one.

virus etiology of cancer hypothesis as the knowledge gained from the study of bacterial viruses is insinuating itself into the pattern of thought even of clinical pathologists.

Dr. Rous, in the ninth decade of his life, has a philosophical attitude about his struggles to gain acceptance for his hypothesis. He feels that this was inevitable, since it takes at least ten years for any new idea to win recognition.

At the present time there are two hypotheses on the cause of cancer which are accorded serious consideration. One of these, the somatic mutation theory, holds that cancer is caused by a local mutation in the somatic, or body, cells. The other hypothesis pins the blame on viruses.

No experimental biological scientist ever puts his complete trust into just one working hypothesis as the ultimate mechanism of a phenomenon. But, even though we have not been able to demonstrate a virus as a cause in human cancer, the line is worth pursuing. Our lack of success up to now is not necessarily a valid argument against such a hypothesis. The known cancer-producing viruses show astounding specificity. The mouse mammary carcinoma virus is infectious only in the mammary tissue of one strain of mice; the leopard frog cancer virus will reproduce—and cause cancer—only in the kidney of the leopard frog. It is becoming increasingly evident that a human being is a walking incubator, nurturing a vast variety of viruses. Those that already have been typed run into scores. Any one of these multitude of viruses might, under optimum conditions for it, be the inciting agent for the shunt to the profoundly altered growths that are clinically classified as cancer. In addition to the known free viruses there is the possibility of the existence in humans of the counterparts of lysogenic viruses. The presence of at least one species of such a virus is well established. The virus of herpes simplex can persist for a lifetime on the nose or lip of a subject, most of the time in an apparently benign symbiosis. However, under some conditions of stress, such as excessive exposure to the sun, some mechanical irritation, or by eating certain allergens, the whole area flares up and exhibits the distressing symptoms of a "cold sore." What other so-called occult viruses we may be harboring we do not know. It may be significant that ultraviolet irradiation which induces virus formation in lysogenic microorganisms is known—in large doses—to incite skin cancer in mice and men.

For an insight into the possible mechanism of cancer production by a virus we can call upon our experience with the bacterial viruses. An invading bacteriophage is known to have the capacity to disrupt the metabolic harmony of the captured cell. It can actually introduce the capacity for the shaping of new enzymes where no such enzymes existed before. A virus may produce cancer by shunting the metabolism of an animal cell into more primitive pathways which defy the usual regulatory restraints; or, a virus may merely interfere with the feedback control mechanisms which normally regulate the relative amounts of various tissue components which are to be manufactured by the cell's machinery.

Whatever the eventual resolution of the viral etiology hypothesis of human cancer may be, it is valuable if for no other reason than its dynamism. Nothing much can be done experimentally with the somatic mutation hypothesis: it is impossible to explore a random, unpredictable event. The virus hypothesis, on the other hand, invites experimental forays which may not lead directly to the eventual goal, the understanding and mastery of cancer, but will certainly yield new information out of which a new pattern and a new path may arise.

The Hungarian biochemist Dr. Albert Szent-Györgyi illustrates the need for a guiding theory, however tenuous, with a story out of his experience in the First World War. He was with the Hungarian army on the Italian front when they sent out a small patrol to reconnoiter the mountainous area in front of them. The patrol did not return and was given up for lost. However, a couple of days later the exhausted and bedraggled patrol crept back into camp. When the soldiers were questioned they said they had become lost in the mountains, but they decided to follow a map which indeed brought them back. The story sounded implausible to the interrogating officer who doubted that the peasant boys who made up the patrol could read a map. So he asked to see it. The soldiers did produce the map which they had followed, but it was of an entirely different area. The virus theory is the best map we have at the present time, for it leads us on to experimental trails.

*But where shall wisdom be found and where is
the place of understanding.*   JOB 28.12

12 . THE BRAIN

CELLS THAT THINK

THE IMPACT OF A GENIUS is sometimes detrimental to a field of research. Freud was such an unwitting brake on the development of brain physiology and chemistry. With brilliant insight he penetrated slightly the mists shrouding the abyss of the human mind; his influence on the therapy of mental ills, as well as on our literature and art—not to mention our dinner conversation—is profound. Yet, despite the bold strides with which he led his disciples in the verbal exploration of the human mind, he had an almost paralyzing influence in another area: he made us forget that the brain is an organ.

It is an organ crammed with chemical machinery. It burns foods; it consumes energy; it builds up and breaks down its own tissues. On top of all this, or rather, with the aid of all this, it performs its supreme function: it thinks. The brain is thus an organ of staggering complexity, but it is an organ, no less than the kidney and the liver are organs. Our complaint is that under Freud's hypnotic influence much effort has been focused on the psychoanalysis of the brain and little on its chemical and physiological analysis.

The implication is not that Freud's influence has been bad. There are any number of puny people who try to eke out a bit of fame by associating their names with the great, if only as their detractors. Attack gets more attention than adulation. The writer has no desire to join the ranks of the fifth raters who would achieve fourth-rate stature by attacking a first-rate mind and its works. Indeed, he believes that with no amount of chemical or physiological research using our current, crude techniques could we have advanced as rapidly in the brief space of a few decades as we did following Freud. The two different approaches should have been encouraged to proceed side by side; psychoanalysis for its rapidly reached immediate gains and physiological research for its slow, ponderous, but more profound exploration of the mechanism of the mind.

Of course, in all fairness, it should be pointed out that there were formidable obstacles to the physical exploration of the brain long before Freud. Until the seventeenth century man's soul and mind were considered by many to be akin to a gas. But then, as we began to unveil the mysteries of gases, refuge was taken behind more impenetrable veils of mysticism; Descartes separated mind and matter. He staked us out into two tightly fenced areas: *l'âme raisonnable* and the *machine de terre*. He conceived a unique metabolism for the brain. The blood was supposed to induce in the brain "a very subtle air or wind, called animal spirits." The dualism of mind and matter dominated our thinking for centuries. Indeed, it still dominates it today.

That the brain might have an independent metabolism, instead of being a Cartesian windmill, began to be recognized from the work of a remarkable physician who turned to chemistry, John Lewis William Thudichum. Although he was born and educated in Germany, he spent most of his life in England. A physician by training and practice, his major work was in chemistry; author of the classic *Chemical Constitution of the Brain*, he also wrote *The Spirit of Cookery* and *A Treatise on Wines;* a remarkably brilliant researcher, he was at the same time a singer of public concert stature.

Thudichum, who worked furiously at his many interests until his death in 1901, was the founder of brain chemistry. His determinations of the gross components of the brain have not, to date, been bested. His skill at unraveling the chemical structure of many components of the brain is nothing less than awe inspiring.

But interest in the chemistry of the brain lagged after the time of Thudichum. The metabolic functions could not vie for the attention of investigators with the unique functions of the brain. And so we behold the anomalous situation that the most remarkable of all organs received the least attention as an organ. Nevertheless, we learned much about the brain from borrowed knowledge. Nature is a frugal inventor. If a mechanism works well in one organ, that mechanism is bound to be installed in a new one. Sugars were broken down and their sun-born energy was sucked up by cells thousands of years before the specialized cells of the nervous system made their appearance. When, finally, these master cells did arise, they were equipped with the very same furnace for the breakdown of sugars which had been found efficient in the other cells. The metabolism of sugars in the brain therefore was unraveled by studying not brain cells but yeast cells and pigeon liver cells. Thus, borrowing a bit of knowledge here, filling in a bit there, we are slowly rounding out the picture of the brain as a metabolic machine.

Of course, spiritual descendants of Descartes may argue: We grant that the chemist can unravel, slowly, the mechanical functions of the brain; but is there any promise that he can correlate the gross chemical functions and such infinitely subtle entities as thought and personality? Such correlations have been made. Again, these tiny glimpses were made possible not from the work of scientists primarily interested in the brain but from the incidental bits of unexpected information which are often the by-products of explorations in biology. For example, the study of vitamins, or rather of vitamin deficiencies, has produced unexpected revelations on the effects of biochemical deficiencies on that master organ, the brain.

Mental disturbances invariably accompany the physical symptoms in pellagra—one of the acute vitamin deficiencies. The patients are apprehensive, fearful, irritable, and easily aroused to anger. These symptoms vanish from pellagrins when nicotinic acid, whose absence is responsible for the disease, is administered.

Nicotinic acid is not unique in throwing the central nervous system out of gear by its absence from a person's diet. In beriberi, induced by the lack of vitamin B_1, severe neurological symptoms are part and parcel of the disease, too. An artificially produced deficiency of biotin converts normal humans into ready subjects for the psychoanalyst's couch.

What is the connection between these vitamins and the functioning of the brain? The source of energy for the brain is the metabolism of sugar. These three vitamins are coenzymes in the various steps of that intricate metabolism. A lack of vitamins is a lack of coenzymes which, in turn, means crippled enzyme systems. It appears that in these cases of vitamin deficiency, deranged enzyme systems in the power plant of the brain produce temporarily deranged personalities.

Another and different line of evidence on the potent effects of chemicals on the functioning of the brain was accidentally discovered by a Swiss chemist in 1943 in a routine investigation of substances derived from ergot. The chemist accidentally inhaled a small quantity of lysergic acid. He became dizzy and

left for home in a dreamlike state, his mind alarmingly filled
with hallucinatory images. "With my eyes closed fantastic pic-
tures of extraordinary plasticity and intensive coloring seemed
to surge towards me." One millionth of a grain of this drug
can induce transient hallucinations mimicking a psychotic state
in a human.

Whether the stresses on a brain are emotional or chemical,
the consequent symptoms are apparently very similar. It may
be hasty to jump to the conclusion that the seat of both dis-
orders is necessarily in the same mechanism. (Whether the
electric power supply to a radio is erratic or one of its transis-
tors is worn out, distortion results in both cases.) Nevertheless
it is inviting to speculate whether the fragile links in the brain
which part under the two different kinds of stresses may not be
the same. That the healing is so slow in one case and so spectacular in the other is not a valid objection to tracing the dis-
orders to the same faulty mechanism. Emotional shortcomings
are not so easily remedied as biochemical ones.

We know of only one ubiquitous mechanism in living things
—enzymes. The "abnormal personality pattern" which makes
one person succumb to emotional strains which a sturdier one
tosses off may yet be shown to be an abnormal enzyme pattern.

There is strong evidence that schizophrenics have definite
biochemical aberrations from the normal. Under stress, certain
enzyme and hormonal systems in such patients are far more
sluggish than those of normal persons. Moreover, schizophrenics
have a copper-containing enzyme in their blood whose level has
been claimed to be correlated with the intensity of the mental
disturbance in the patient. Indeed, it is also claimed that this
same enzyme can be shown to vary in normal people under the
influence of drugs which can produce transient psychotic states.
The time may yet come when we will probe into a patient's

erratic enzymes rather than into his emotional history. However, at the present rate of progress, enzyme analysts will not
be hanging out their shingles for another couple of hundred
years.

Can we approach a means of studying chemically how our
brain commands our muscles to move or how that remarkable
organ thinks? We have been able to reveal recently a tiny fragment of the chemical machinery which communicates nerve
impulses to muscles.

The wiring for the telegraphic system from brain to muscle
has been known for a long time. The nerve cells have long, thin
threads—some of them several feet long—leading from the
spinal cord to the various muscles. Messages are sent along this
network—at a speed of about 120 feet per second—ordering
the muscles to perform their functions. But there is a gap between the muscle and the end of the nerve fiber; there is no
contact at all between the two types of cells. How, then, is a
message sent across the gap? What is the messenger which
hands over the telegram to the obedient muscle?

Some chemicals, for example, adrenalin, when applied to
muscles, make them behave as if they were stimulated by the
nerves leading to them. It had been conjectured that perhaps
the nerve impulse is conveyed to the muscle by the shooting
of some such potent chemical into the space between nerve
and muscle.

The conjecture was enshrined as a fact of physiology by a
very simple experiment devised by the Austrian pharmacologist
Dr. Otto Loewi.[1]

The experiment is remarkable not only for its neat simplicity
but also for the circumstances of the birth of the idea: Dr.

---

[1] Pharmacology is still another branch of experimental biology. Pharma-
cologists study the effect of drugs on living things.

Loewi dreamed it. He tells that he had a vivid dream in which he performed the crucial experiment. But, by the next morning he had forgotten the details of the fruitful dream. Fortunately, however, he dreamed it again. This time Loewi woke up and jotted down the idea. Then he performed the dream-borne experiment which, eventually, also fulfilled a dream: he received the Nobel Prize in medicine for it.

An organ which, like the heart, operates involuntarily has two sets of nerves leading to it; one to stimulate it and one to inhibit or retard it. Thus our brain keeps such an organ under control with two reins.

Loewi exposed the heart of a frog and severed the nerves leading to it. He then stimulated with an electric shock the heart of another frog with intact nerves. The normal result of such a stimulation is a decrease in the frequency of the heartbeat. He then withdrew some blood from this stimulated heart and placed it in the first denerved heart. But nothing happened.

He repeated the very same experiment but he replaced the blood in the heart having the intact nerves with a salt solution. This time when he transplanted the salt solution from the stimulated heart into the nerveless one, something did happen. The heart which could receive no stimulus from its severed nerves behaved on receiving the salt solution as if it had been stimulated by a nerve: its heartbeat decreased. In other words, the stimulated heart had something released into it which could be transferred and was still potent enough to stimulate another heart. This was the first proof that there is a messenger which jumps the gap between the terminal of the nerve telegraphic system and the addressed muscle cells.

Later, Sir Henry Dale, the British physiologist, proved the identity of the messenger. It turned out to be a relatively simple chemical called acetylcholine, a substance which had been

messages are sent from brain by release of acetylcholine

made in the laboratory years before. Once again, the random product of the organic chemist's skill turned out to be a product of the cell as well. And, also, once again the answer to one question raised other equally baffling questions: If messages are sent by the release of a chemical, how can the messages be repeated at frequent intervals? Why does not the first message persist through the continued action of the acetylcholine?

The answer is that there is an enzyme present in blood which rips acetylcholine into impotent fragments. Thus, once a message is delivered to the muscle cells this enzyme tears up the messenger and the stage is set for a new communication. This is the reason for the failure of Loewi's classic experiment until he replaced the blood in the frog's heart with a salt solution. By the time the blood was transferred from heart to heart, the enzyme had completed the destruction of the messenger. (Incidentally, the German nerve gas DFP acts by inhibiting this very enzyme. Once the enzyme is rendered impotent by DFP, no new messages can be sent to the muscle and the victim becomes paralyzed.)

The work on acetylcholine is testimony to the fact that we can approach with the tools of chemistry and physiology one of the tiny segments of the machinery of the nervous system. But can we at present approach a similar study of the profound functions of the brain? Can we ever answer in chemical terms the ancient plea of Job: "But where shall wisdom be found and where is the place of understanding?"

Biochemists are confident that some day the molecular site of understanding will be found and that its chemical machinery will be mastered. Unfortunately this confidence springs from the contemplation, not of the current progress in this field, but of history. Many times in the past have scientists, bent on mechanical interpretations of the phenomena of life, encoun-

tered forbidding signs: "Thus far and no further." The signs
have faded; the barriers between the nonliving and living worlds
have crumbled. One by one, every part of life's machinery
proved to be physicochemical machinery.

Since we know today that the source of energy for the brain
is chemical and that it relays its messages through physicochemical means, it is almost an article of faith that some day we
shall find that memory, thought, and will are molecular mechanisms as well.

The metabolism of foods, the release of energy and its conversion into motion, even the propagation of nerve impulses,
all this we can explain in chemical terms today. But how does
a mass of tissue made of water, some fats, and proteins have a
memory? How does it think? How does it write a symphony?
Of this we know nothing.

What is memory? Is it carved into the structure of a molecular complex? How is it perpetuated year after year? We know
that the brain, like other tissues, is in a state of flux. It is constantly broken down and rebuilt. How does a memory remain
intact, sometimes for a lifetime, amidst this constant destruction and repair of tissue? Are the molecules which house a
memory rebuilt constantly as the antibodies are? Of this, too,
we know nothing. This is the depth of our ignorance about the
machinery of life.

While we lack real knowledge about the machinery of the
mind we are not lacking in conjectures about it; the absence
of one fact creates a vacuum which easily accommodates a score
of conjectures. The current vogue is to find analogies between
the brain and an electronic computing machine. Our brain is
supposed to be just one vast Univac machine containing fifteen
billion valves—or cells. This kind of reverse anthropomorphism
—or is it machinomorphism?—is contrived and lacks justifica-

tion based on tenable evidence. To be sure every known mechanism of every living organism operates within the bounds of the laws of thermodynamics and mechanics. In this sense every living thing is a machine. But the machine was not made in the image of some handicraft of man.

It is belittling of the grandeur of life's machinery and of the inventiveness of nature to assume that after a billion years of evolutionary experiments nothing more ingenious had evolved than a replica of some invention of man. We have no idea at the present time what the molecular mechanism of the mind is and it is of little value to cover this chasm of ignorance with a flimsy drape of contrived analogies. Knowing what we do not know is the first step to knowledge.

It is surprising, in view of this impenetrable blanket of ignorance which at present hides completely the mechanism of thought and the shaping of a personality, how ready are some scientists to proclaim at the end of their careers that they have found the ultimate truth in the bottom of their test tubes. In our present state of knowledge, or rather of ignorance, it is presumptuous for scientists to turn into prophets on the basis of their scientific experience. While the tools and methods of science have borne impressive fruits, the area in which those tools can be wielded at present is quite limited.

Scientists should leave behind their mantles of authority when they abandon the realms explored or explorable by science.

The biochemist can make very thin slices of various organs and can detect enzyme mechanisms in the surviving cells. He can slice liver 0.25 millimeters thick and learn that it makes cholesterol. The butcher slices liver 25 millimeters thick and also knows some of its dietary potencies. The scientist's slices are a hundredfold thinner; his knowledge of the functions of the liver is perhaps a hundredfold more detailed. But is he therefore a hundred times better qualified to ponder the infinity of space, the endlessness of time, the origin of matter-energy?

It is odd how readily a few scientists abandon life-long habits of buttressed reasoning and cautious utterance once they leave their circumscribed fields and take a fling in the wider realms of mysticism. For example, the distinguished physical chemist, the late Pierre Lecomte du Noüy wrote in *The Road to Reason*:

There is an element in the great mystics, the saints, the prophets, whose influence has been felt for centuries, which escapes mere intelligence. We do not admit physical miracles, because they are outside the actual framework of our knowledge; yet we admit the reality of Joan of Arc, who represents a real and confounding miracle.

This is a fallacy of partial truth. Of course Saint Joan is a real and confounding miracle. But so was her lowest yeoman a miracle, or, indeed, so was the horse she rode. The miracle is not a specific life. The miracle is *any* life!

The poet, not unexpectedly, is more sensitive to the mystery and splendor of all things living. Wordsworth wrote:

> To me the meanest flower that blows can give
> Thoughts that do often lie too deep for tears.

There are some scientists who at the end of their careers enumerate all that is still unknown and, perhaps, unknowable. On the basis of the enormous gaps in our knowledge they exhort us to faith. Ignorance of natural phenomena is an unsteady pillar for the edifice of faith. It is an ephemeral stanchion at best. The mystery of yesterday is the commonplace of today; the unknown of now will be explored tomorrow. Three hundred years ago the mechanism of fire was just as baffling as the workings of the human mind still are today. Should men have been exhorted to faith in those days on the basis of the mystic wonder of a fire?

Pasteur, that greatest of biochemists, who pioneered so much in the physical realms, pre-empted the role of scientist-mystic as well. He wrote in his speech of acceptance to the French Academy:

What is beyond? the human mind actuated by an invincible force, will never cease to ask itself: What is beyond? . . . It is of no use to answer: Beyond is limitless space, limitless time or limitless grandeur; no one understands those words. He who proclaims the existence of the Infinite—and none can avoid it—accumulates in that affirmation more of the supernatural than is to be found in all the miracles of all the religions; for the notion of the Infinite presents that double character that it forces itself upon us and yet is incomprehensible. When this notion seizes upon our understanding, we can but kneel. . . . I see everywhere the inevitable expression of the Infinite in the world; through it the supernatural is at the bottom of every heart. The idea of God is a form of the idea of the Infinite. As long as the mystery of the Infinite weighs on human thought, temples will be erected for the worship of the Infinite, whether God is called Brahma, Allah, Jehova or Jesus; and on the pavement of those temples, men will be seen kneeling, prostrated, annihilated in the thought of the Infinite.[2]

It seems to the writer that the contemporary mystic scientists have added little to this in substance, or, for that matter, in style.

[2] René Vallery-Radot, *The Life of Pasteur* (New York, Doubleday, 1916), p. 342.

The fairy tales of Science,
and the long result of time.
TENNYSON

13. ATOMS INTO LIFE

THE International Geophysical Year—yielding that stunning achievement of physics, the man-hurled satellite—has turned our gaze skyward. So much so that some of us are getting a crick in the neck. It is beginning to sound as if we were a nation of Jules Vernes with unlimited resources and manpower. We are being exhorted from all sides to emulate Stephen Leacock's hero and jump on our rockets and ride off in all directions. The fare for these jaunts is commensurate with the distances: they are both astronomical.

The fear of many of us is that while the harvest from such prodigious efforts may be chimerical they will certainly serve to deter progress in the earthbound sciences. For not only would preparation for space travel be a vast drain on our material resources but, far more damaging, it would siphon away young men and women who might otherwise take up the challenge of more fruitful scientific problems.

However, the recent successes of rocketry—which have elevated interplanetary travel from the impossible to the improbable—do offer a promise, however tenuous, which may make them worth all the wealth which has been lavished on them. If we could but land a biochemist on Mars to search for living forms there, many of us would consider that the rockets have redeemed themselves. For a study of the living forms which may be eons behind us in their evolutionary time schedule may contribute to our understanding of our own living mechanisms. Moreover, if we are lucky we may encounter primitive forms at the dawn of life which have long ago vanished without a trace from our own planet. Such a prototype of life, if subjected to a chemical dissection, might serve as the Rosetta Stone for an understanding both of the mechanism and the origin of life.

At the present time, without such a model in front of us, very little can be said about the origin of life that does not evaporate when the Klieg lights of critical scientific inquiry are turned upon it.

Apart from Divine Creation, there have been two different suggestions on the origin of life on Earth. The first one was proposed, among others, by the distinguished chemist Arrhenius, whose enduring contribution was the recognition that matter can exist in the form of electrically charged particles. He coined the term *ions* for such particles. In 1908 Arrhenius suggested that life on Earth started from some spores which drifted here from interplanetary space.

The propelling force for the interplanetary or interstellar space voyages was supposed to be the pressure exerted by light rays. Arrhenius calculated that a bacterial spore could move with great speed in the interstellar voids. He estimated that one could reach us from Alpha Centauri in 9,000 years. Such a suggestion, of course, evades the issue: it is nothing more than buck-passing on an astronomical scale. Moreover, we now know from our rocket probings of outer space that Earth is surrounded by a zone rich in ions of high energy, the so-called Van Allen belt. The passage of a spore in a viable state through this zone, unprotected by a lead shield, is unlikely. The Arrhenius theory on the origin of life is thus riddled by the very particles he named. Moreover, it is difficult to see how the spores, without ceramic nose cones, could survive the frictional heat once they were in contact with the atmosphere.

Of the indigenous origin of life on Earth there are two different theories that need to be seriously considered: The first one holds that the primordial living forms were autotrophic, that is, they could synthesize all of the components of their structure from the inorganic substances available on the surface of Earth, just as our contemporary green plants do.

Such a living form would have to arise, equipped with arsenals of integrated enzymes to perform the intricate tasks of weaving atoms into substances which theretofore did not exist. The probability of the chance occurrence of such an organism is so small that this theory in effect invokes Divine Creation. From the point of view of an experimental scientist a theory which depends on a miraculous event which occurred eons ago is sterile: It is beyond our experimental reach, there is nothing we can do with it beyond stating it.

The second theory postulates that the primordial forms were merely accretions of complex organic compounds which we presume abounded in the oceans of the primitive Earth. These pre-living forms merely agglomerated the many compounds in the laden seas and only eons later did they acquire synthetic abilities. According to this hypothesis the autotrophs, organisms with complete synthetic capacity, were the result of progressive chemical evolution. This conjecture is the most persuasive for the experimental biochemist.

The idea of spontaneous generation of highly complex contemporary organisms was widely held until the end of the seventeenth century. Worms and maggots were supposed to spring fully formed, if not from the head of Zeus, like Minerva, then at least from the putrefying heads of slaughtered cattle.

The Tuscan physician and poet, Francesco Redi, proved in 1668, by well designed experiments, that the maggots in rotting meat originated from eggs deposited by flies. His conclusions were attacked by his contemporaries, oddly enough, on the basis that they conflicted with the Scriptures. A hundred years later (1765) the Abbé Spallanzani had to refute spontaneous generation again, this time of microorganisms. He showed that microorganisms did not arise in meat and vegetable infusions which were boiled and sealed. His conclusions were also attacked. His findings were scuttled by the rationalization that he destroyed vital forces by the vigorous boiling. The notion of the spontaneous generation of contemporary complex organisms was not laid to rest until Pasteur performed his ingeniously designed experiments in the latter half of the nineteenth century.

The years 1857 to 1859 were bountiful years in our cultural history. Pasteur published his incisive studies on fermentation in 1857. This not only led to a refutation of the false theory of spontaneous generation, but also started us on the path leading to an understanding of the mechanism of life as an integrated series of physicochemical reactions catalyzed by enzymes. In 1858 and 1859 Wallace and Darwin published the theory of evolution by natural selection. This brilliant insight into the workings of nature suddenly revealed the road which the complex modern creatures, be it man or mole, must have traveled in their ascent from some primordial living prototype. In the hundred years since, biochemists have accumulated irrefutable evidence at the molecular level for the common origin of contemporary creatures: The mighty whale swings his tail with the aid of adenosine triphosphate; the same compound of intricate structure also serves as life's battery for the quaint little microorganism *Thiobacillus thiooxidans*, which makes a living out of oxidizing sulfur to sulfuric acid. The chance occurrence of this and of other compounds for the same function in these two organisms without assuming common ancestry is of the same order of improbability as finding on a distant planet or another solar system a copy of the Taj Mahal identical to the last stone and also housing the remains of some potentate.

The theory, or better, the law of evolution through natural selection eliminated the need to consider the origin of life for each species. But it insistently forced on our minds the problem of the origin of the first ancestral living forms; for the law of evolution is a law of dynamic change, not of static origin.

The idea of the spontaneous generation of some prototype living form from the accumulated chemicals on the cooling primitive Earth became part of the *Zeitgeist* of the latter part of the nineteenth century. The German philosopher Ernst Haeckel wrote in 1868 in *The History of Creation* that the first living or pseudo-living forms must have been merely "homogenous, structureless, amorphous lumps of protein" originating from dissolved matter in the primeval seas. That Darwin himself entertained such ideas is obvious from a letter he wrote in 1871. In it he defended the idea of such primeval creation against the objection of the lack of repetition of it at the present time.

It is often said [he wrote] that all the conditions for the first production of a living organism are now present, which could ever have been present. But if (and oh! what a big if!) we could conceive in a warm little pond, with all sorts of ammonia and phosphoric salts, light, heat, electricity, etc., present, that a protein was chemically formed ready to undergo still more complex changes, at the present day such matter would be instantly devoured or absorbed, which would not have been the case before living creatures were formed.

It is interesting that both Haeckel and Darwin speak of proteins as the key structure of their primeval forms some twenty

years before enzymes were discovered and almost sixty years before enzymes were proved to be proteins.

Implicit in this theory of the agglomeration of chemicals to
form protein is the presence in the primeval seas of the component parts of proteins, the amino acids. There were several
attempts to test the possible synthesis of amino acids under the
conditions which may have existed on the primitive Earth.
Mixtures of gases, presumably simulating those that abounded
on the cooling Earth, were subjected to electric sparking or to
ultraviolet irradiation and were analyzed for the presence of
sugars and of amino acids. The results were unimpressive. The
analytical methods were so poor until recently that the claims
for the synthesis of substances other than sugars under the conditions reported were not convincing. Moreover, all of the investigators made the wrong assumption, that carbon had been
present in the primitive Earth's atmosphere in the form of
carbon dioxide.

In 1936 a Russian biochemist, Academician A. I. Oparin,
published a fascinating book, *The Origin of Life*. This has since
become the source book for any discussion in this area which
tries to be rooted not in fantasy, but in the few available facts.

In the first place, Oparin concluded that carbon must have
been present in the primitive atmosphere of Earth, not as carbon dioxide but in the form of hydrocarbons. These are compounds made of chains of carbon festooned with hydrogen
atoms. Hydrocarbons can be the sources for a large variety of
different organic compounds—alcohols, acids, aldehydes—some
of which could react with ammonia to form amino acids.

But a still more plausible nonliving source of amino acids was
soon to be demonstrated. Dr. Harold C. Urey, on returning to
academic life after the Second World War, left the field of
isotopes and took up, of all things, the study of the origin of

the planets. He has since made contributions equally as brilliant as his earlier works on the concentration of isotopes. Dr. Urey, though completely unaware of Oparin's work, calculated that the stable form of carbon in the presence of excess hydrogen on the primitive Earth would be methane. This is the most primitive hydrocarbon, consisting of one atom of carbon to which are attached four atoms of hydrogen. Similarly, nitrogen would be most likely present in the form of ammonia, which is composed of an atom of nitrogen with three atoms of hydrogen joined to it. This too was a reaffirmation of Oparin's conclusion.

Confirmation of the Oparin-Urey hypothesis on the nature of the primitive atmosphere of the planets is provided by astrophysics. From a study of their spectra we are certain that the atmosphere of the giant planets Jupiter, Saturn, and Uranus still consist of methane and ammonia. The reasons for their primitive state are threefold. Their temperature is much lower than of Earth, and therefore the chemical reactions are slower; since they are far away from the sun, the ultraviolet energy falling on them is less intense; and, finally, since their mass is greater, their gravitational pull is greater and thus they could keep their original atmosphere captive for a longer time.

If methane rather than carbon dioxide was the primeval link from which the chains of organic molecules were forged, then the reason for the failure to produce amino acids in attempts to synthesize these compounds under the erroneously presumed conditions becomes clear. Even the best chef's soufflé will not rise if he uses the yolk instead of the white of the egg.

Dr. Urey wrote in 1952:

It seems to me that experimentation on the production of organic compounds from water and methane in the presence of ultraviolet light of approximately the spectral distribution estimated for sunlight would be most profitable. The investigation of possible effects of electric discharges on the reaction should also be tried since electric storms in the reducing atmosphere can be postulated reasonably.

Such experiments indeed did prove profitable. The passing of electric sparks through a reconstructed primitive Earth atmosphere was worked out by one of Dr. Urey's young graduate students, Dr. Stanley L. Miller. Such an experiment was carried out by him somewhat as follows. Into an appropriately constructed instrument were introduced water and the three gases, methane, ammonia, and hydrogen. The glass apparatus was so designed that the water could be boiled to a vapor, then condensed to a liquid, and electric sparks could be continuously passed through the area where water vapor and the three gases were passing. The cooling and sparking in the absence of oxygen was continued for as long as a week. At the end of this time the reaction was stopped, the vessel was cooled, and it was opened to allow the gases to escape. Then, after the water was evaporated, came the exciting moment. There was a solid residue. If there were no reaction there would have been no residue, for initially there were no solids in the system.

The analysis of the newly created solids was a very simple procedure by paper chromatography, the ingenious method for the analysis of minute amounts of chemicals which we discussed earlier. The solids turned out to be a mixture of a variety of organic compounds, including several amino acids in large amounts. This was exciting news, but more recently even more satisfying findings were reported by another young biochemist. If to the same mixture of gases some iron sulfide is added, and it is then irradiated by ultraviolet light, some of the other amino acids such as phenylalanine and the sulfur-containing methionine are produced, as well as complex aggregates of these and other amino acids. The synthesis of these complexes,

or polypeptides, is of particular interest, for they are held together by the same binding system which builds protein molecules. Since polypeptides are assembled this way, the weaving
of still more amino acids into a larger protein molecule may
be just a matter of time. Any system which can weave a napkin
can also turn out a tablecloth.

These are extraordinarily suggestive discoveries. The structure
of phenylalanine is rather intricate. It consists of a ring of six
carbon atoms, to five of which single hydrogen atoms are attached. From the sixth carbon stems an appendage of three
carbon atoms. To the second of these carbons is attached a
nitrogen plus two hydrogen atoms, or the amino group; to the
third carbon are attached two oxygens and a hydrogen, the
acidic group.

The alignment of these twenty-three atoms of four different
atomic species into the exact pattern of one of our essential
amino acids under the influence of ultraviolet irradiation appears to be almost magical. And indeed it is: the magic of
molecular reactivity. To the chemist, however, such molecular
transformations induced by heat, pressure, and irradiations are
less than magical; they are the source of his daily bread. The
most important step in making that wonder drug aspirin [1] is to
heat together, under pressure, carbolic acid and carbon dioxide
—presto! out comes salicylic acid.

How is that particular substance, phenylalanine, formed under
the influence of ultraviolet light? The fact is that there are
literally hundreds of other compounds formed under those conditions. But most of these are too unstable to exist for more

[1] Since aspirin was a gift of the chemists in an era of less flamboyant slo-
gans, and since we have learned to take it for granted through generations of headaches, it has never received the adulatory name it deserves:
wonder drug. But it is just that. Like penicillin, aspirin achieves its mission without harmful side effects. As to their relative merits, penicillin
merely saves our life; aspirin often makes it tolerable.

than fractions of a second. Phenylalanine and the other amino acids are stable, condensed systems of the atoms involved, and once they are formed they endure, we know from work on fossils, for as long as three hundred million years. Thus they will accumulate at the expense of the other unstable transient forms.

A surprising aspect of the amino acids made under these conditions is that they are almost uniformly the ones with the amino groups on the second carbon atom. These are the so-called *alpha* amino acids. They are the ones our body's proteins, as well as the proteins of fossils, are made of. Had the tendency of carbon atoms been different, say to put the amino group on the third carbon, that is, to form *beta* amino acids, our proteins would have had entirely different properties; we would probably be entirely different looking creatures. But this is only one aspect of the life-shaping power of the carbon atom. The promise of life itself was locked into the structure of the carbon atom the moment it came into being. For, of all of the elements known, carbon has the greatest capacity for uniting with itself into molecules of vast size and infinite variety. Only from so versatile an element, which has the ability to rearrange itself into an infinity of patterns, each with new and unique properties, could so wondrously complex a system as life arise. It has been suggested that life was created the moment matter was created. This may be true in the sense that the spark of life started flickering in the valence electrons of the carbon atom.

How did life arise? I ought to warn the reader that the only facts known are those I have outlined above: amino acids and other organic molecules could be formed with ease on the primitive Earth. The next epoch we can document is the Cambrian period, the time of the oldest remnant fossils. What happened in the billions of years in between is, at the present time, almost anybody's guess. Conjecturing on it is a game everybody can play and, indeed, almost everybody does. It is a pleasant game for many of us. A career in science is often a career of intense preoccupation with minutiae. It is a joy from time to time to escape our narrow corner, cast off the restrictions imposed on us by the methods and the ethics of science, not to mention the editors of science, and assume the role of cosmologists. Speculation on events which occurred eons ago under conditions which cannot be repeated is an engaging pastime, one that is nowhere nearly so taxing as establishing a tiny little fact about our contemporary world which may be promptly challenged by a contemporary colleague.

If the reader would like to come and have coffee with me, here's the way one speculative monologue on the origin of life might go.

Let us visualize the face of the primitive Earth. There are many lines of evidence indicating that at one time it was a seething hot mass which slowly cooled to a temperature where the enveloping clouds of steam could condense to form the vast seas. The atmosphere, say about three billion years ago, consisted essentially of methane, ammonia, hydrogen, and water vapor. Enormous amounts of energy cascaded on this mixture of gases in the form of sunlight (at the present time this rate is 260,000 calories per square centimeter per year). Lightning, from electrical storms, also provided energy, albeit most likely at a much lower level. (At the present time this rate is only about 4 calories per square centimeter per year.)

Prodigious amounts of organic compounds were continuously being produced in the atmosphere. As fast as they were shaped, these compounds cascaded below with the frequent torrential rains to accumulate in the seas below. If all the carbon which is estimated to be present on the surface of Earth today were in such soluble organic form, the seas might have been laden with two billion billion (2×10^{18}) tons of material. J. B. S. Haldane, the English biologist, who is one of those scientists who speculates in print on the origin of life, called these compound-laden seas a "hot thin soup." If all the carbon was really present in the form of organic compounds, especially as polypeptides, it was quite a "thick soup." [2] (These speculations do not deserve more precise units of measurement than "thick" and "thin.")

Whatever one's preference for soup, be it bouillion or mulligatawney, the primeval seas undoubtedly abounded in the organic compounds which could serve as the building blocks of eventual life. What particular substances may have been present in those primeval seas?

It is certain that the shaping of organic molecules in the primitive atmosphere was far more prolific than would be indicated by the few experiments which have been performed to probe this area. With a positive finding we are twice blessed: we have not only something new, but a rich promise of more findings to come; with a negative finding we are twice vexed: a reaction may just not have gone under one set of conditions, or it may never go. If a chemist reports that he found no evidence for the synthesis of purines or porphyrins after exposing the presumed primordial gases to sparking or to irradiations for a few days, he is really saying that under one set of conditions he has not found them in concentrations high enough for the assay methods he used. Our most sensitive tests will not detect these substances in amounts below a ten-millionth of a gram

[2] There is a danger in quoting references in this area. It adds too much of the authority of the apparatus of a scientific article. Normally when we give a reference it is to a *fact* reported by someone and usually confirmed by others. In this area, however, a reference often merely cites someone else's conjecture.

(10^{-7} g.). The number of molecules which can still elude the prying chemist is astronomical. It is on the order of a hundred thousand billion (10^{14}) molecules.

The chemist runs his experiments for days, in tiny volumes. What might have accumulated in the primitive oceans in a few hundred million years is beyond our imaginings. Some of the silicate mineral beds along the shores could have selectively concentrated the rarest of organic molecules.

It is probably significant that the structural units which are needed for the assembly of life are extraordinarily rugged compounds. The stability of the amino acids has already been discussed. Equally rugged are the porphyrins. These molecules serve as the cutting edge of several pivotal catalytic molecules: they are present in chlorophyll, where they are involved in photosynthesis; in hemoglobin they transport oxygen; in the cytochromes they perform cellular respiration. Coproporphyrin is an excreted end product of porphyrins resulting from the death of red cells. Coproporphyrin is so stable it has been found intact in the fossilized excreta of crocodiles which may have been basking in the sun 50 million years ago. The purines and pyrimidines which are the building blocks for the nucleic acids are also very stable molecular structures. It would appear that life evolved out of molecules which could sit around in the primeval seas for years, waiting to be picked up, without becoming shopworn.

How then did life arise from this fluid warehouse of stable organic molecules? We are under a profound handicap when we come to grips with this problem: we have no models to look at! The biological scientist can but observe life's mechanism, he can rarely predict them. The one certainty that emerges from a study of the history of the biological sciences is that nature's methods have invariably outstripped our imagination.

Could the human mind have conceived that we have ascended from some notochord-bearing little fish like a lancelet or an eel, without the evidence in the rocks staring us in the face? Could we have visualized a priori the living machine driven on phosphate bond energy?

Unfortunately, there will be no models of pre-living forms to look at until possibly some day a well-trained biochemist may be landed on Mars. And, alas, we may be too late even there, for that planet's oceans have evaporated. If the changes in color with the changing seasons indicate vegetation, as they do to some experts, then Mars has quite an advanced form of life.

Meanwhile we must be satisfied with conjectures, for we cannot dignify our speculations by assigning to them the well-defined term in the semantics of science, hypothesis. A hypothesis should be an integration of observed phenomena, and it must predict other phenomena, either existing or wrought by further experiments. How can we design an experiment to repeat an event which may have been the culmination of perhaps a billion years of molecular interactions?

Every conjecture on the origin of life which has been put forth is essentially an extrapolation of some of the known mechanisms in contemporary living cells. We merely endow some primordial protein molecule which was shaped through the inexorable laws of permutation and combination with some of the known properties of today's protein molecules. For example, we speculate that photosynthetic mechanism could fumblingly start when some clump of protein molecules, endowed by their random structure with catalytic properties, was tossed by a wave onto a magnesium silicate deposit encrusted with porphyrin molecules. For these are the essential ingredients of a contemporary photosynthetic system: porphyrin, magnesium, and protein enzymes. By extrapolation of some of the known laws of microbial genetics we ascribe a survival potential to such a primordial clump. Other protein clumps, without synthetic capacity, would exhaust some of their nutrients and would be doomed to disintegration, whereas the clumps with some synthetic capacity could survive a temporary local famine. The more synthetic potencies a clump would have the greater would be its chance of survival. Thus, natural selection could operate at the molecular level, and it would favor the survival of aggregates with increasingly complex synthetic capacity and structure.

Such speculations are stimulating, but, it appears to the writer, scientifically not fruitful; a speculation which can lead to no experiment which could increase our arsenal of knowledge is barren.

What can we know about these pre-living forms which could leave no traces in the ancient seas. They may have been microscopic, or as large as contemporary jelly fish, or they may have been vast blobs the size of a continent. It is enough for the present to assume that just as under a given set of circumstances the most complex of amino acids are assembled from small prototype molecules with the energy of sunlight in a brief time, so in a billion years a living, self-duplicating form might have arisen by molecular evolution from the precursor building blocks abounding in the primeval seas. Just as an amino acid can accumulate because of its stability at the expense of less stable molecular configurations, so could such living forms be selected from the billions of random experiments of nature through their own unique aptitude for survival: the ability for self-duplication.

If such molecular evolution seems implausible, so does organic evolution, despite the evidence written in the rocks, seem implausible. It is easier for the writer to believe that living unicellular organisms arose after a billion years from random molecular experiments of nature than that Beethoven and Einstein arose from some gasping little eel.

A materialist theory of the origin of life from preformed molecules is persuasive because it assumes a gradual evolution starting at the molecular level. Evolutionary transformation is a pervasive pattern in our universe. Nature, that arch conservative, abhors abrupt change. The atoms have evolved from prototype particles, the lightest atoms were packed into heavier ones; the solar system repeated on a gigantic scale the structural pattern of the atom. Gradual evolution has certainly been the ruling pattern of life in the past couple of hundred million years. Molecular evolution may have preceded this by a couple of billion years.

It is only the first stage of this theory, the aggregation of random protein molecules into a self-perpetuating system, which is highly improbable. After that the theory, with its life-evolving progression, gains in probability.

But, inherent in every description of creation is an increasing improbability as ultimate origins are approached. The creation of Adam was a miracle second only to the miracle of the creation of the Universe. But the shaping of Eve out of a living rib was merely a bravura feat of tissue culture.

For I dipt into the future, far as human eye could see
Saw the Vision of the world, and all the wonder that would be.

TENNYSON

14. THE BEST IS YET TO BE

"BIOLOGY began as it will end—as applied chemistry and physics," wrote Dr. Hans Zinsser, the eminent bacteriologist, in *Rats, Lice, and History.*

It is good to have this prophecy from an outstanding classical biologist, for, if it came from a biochemist such a statement might sound like the irresponsible babbling of a partisan enthusiast.

The classical biologist, or rather the biologist with the classical techniques for the study of life, is doomed eventually to technological unemployment. He is doomed by the nature of the object of his study. A living cell is a patterned aggregate of ions and molecules. Life is a sequence of interactions between those ions and molecules. It would appear obvious that the most rewarding study of life must be the study of the physicochemical reactions which *are* life. Unfortunately, we have just begun to study life at the molecular level. The road to this stage in the growth of biological science was long and difficult.

The science of gross anatomy, the study of the shape and structure of the organs of the body, began in the sixteenth century. Differences in the structure of the various organs were noted, but the reason for those differences remained obscure until the microscope enlarged human vision. With the microscope the source of those differences was found: each different tissue is composed of cells unique to it. During the nineteenth and the early twentieth centuries the anatomical differences discernible under the microscope were gathered into the body of knowledge called histology or microscopic anatomy. At the same time, the functions of these various tissues were studied and assembled into the discipline called physiology.

Meanwhile the chemist was sharpening his tools and exploring with them our inorganic universe. Finally, when the defeatist taboos were lifted, he dared to undertake the chemical exploration of the living world as well. It was found that the pancreas dominates the metabolism of sugar not because of the unique structure of that organ, but because its cells produce unique molecules, insulin. Thus we arrived at descriptive biochemistry, or, as the author likes to call it, molecular anatomy.

So far the biochemist's efforts have been directed mainly toward the completion of the atlas of molecular anatomy. He has been spectacularly successful in the extraction and identification of the molecular components of the cell. He has catalogued the amino acids, vitamins, hormones, and enzymes. But while all this knowledge is impressive and is of tremendous value in nutrition and in medicine it is astonishingly incomplete knowledge. We know next to nothing about the mode of interaction of those components of the cell. Through the efforts of the biochemist we know the structure of the hormone insulin. That its clinical potency borders on the miraculous is common knowledge. But how a molecule of that particular structure exerts such a profound influence—of this, we know nothing. We must begin to assemble patiently the basic knowledge for still another science, molecular physiology—the study of the relation between molecular structure and cellular function.

The slightest alteration in the chemical architecture of a

component of the cell changes its behavior, altering or abolishing its potency completely. The difference in the structure of linoleic acid, the essential fatty acid whose absence from the diet causes the death of a rat, and oleic acid, which cannot substitute for it, is the deficiency in the linoleic acid of two protons and two electrons. Precisely how the lack of those four minute particles confers life-saving properties on linoleic acid is not known. To ascribe the differences in the potencies of the two fatty acids to the great specificity of some enzyme system into which the indispensable fatty acid must fit is merely a restatement of ignorance in different terms.

This is an area in which future research promises to bring the richest rewards. For we know nothing of the relation between structure and function of the huge pivotal molecules of the cell, the proteins, and the nucleic acids. The riddles of the normal and abnormal functioning of the cell are locked within those structures. The monstrous riddles of the so-called "degenerative diseases"—cancer, multiple sclerosis, arteriosclerosis—are undoubtedly there also.

Through knowledge of descriptive biochemistry the two other large categories of man's diseases—the deficiency diseases and the infectious diseases—have been mastered. The external deficiencies can be remedied by administering vitamins, essential amino acids, and the appropriate minerals. If the deficiency is an internal one, such as an insufficient supply of hormones, the diminished production can, in many cases, be supplemented by extracts of the organs of other animals or by synthetic preparations. The infectious diseases have been all but mastered with the antimetabolites, the antibiotics, and other chemotherapeutic agents. It is but a question of time before drugs effective against the few remaining recalcitrant infectious agents are found.

Against the degenerative diseases, however, science and medicine are still well nigh powerless. Progress against these, it is feared, must await more extensive knowledge of molecular physiology. The tools of the chemist may prove insufficient to this vast task ahead of us. Chemical analysis reveals merely the components of an aggregate of molecules, it can give very little information about the three dimensional structure of the aggregate. It is becoming increasingly apparent that the geometry of a molecule and its relative position to adjacent molecules is as important to its function as is its composition. For the topographical exploration of molecular constellations we shall have to call on the physicists and physical chemists to guide us, for their tools are better suited for the task. Their contributions in this direction are already impressive. A classical biologist discovered the bacteriophage. He could describe it only as an elusive "factor" toxic to bacteria. Then came biochemists who isolated viruses including the bacteriophage. From their chemical analyses we learned that viruses contain nucleic acid and protein. But what is a virus? does it have a structure, or is it merely a molecule in solution? These questions could only be answered after the physicist focused his electron microscope on the elusive bacteriophage. In Figure 3 there is a stunning photograph of a bacteriophage taken with an electron microscope. We must recall that this speck of protoplasm striving toward life is so small that thirty million of them could dance on the head of a pin. The bacteriophage can be disrupted and its various anatomical parts can be separated by subjecting them to varied gravitational forces in an ultrahigh-speed centrifuge which is another instrument of the physicist.

In Figure 4 we see only the stems of the bacteriophage. The round ratchet wheels are stems which were caught by the electron microscope standing upright. From this picture we deduce that the stem of the bacteriophage is a hollow tube—apparently with a corduroy surface for rigidity—through which the DNA is injected into the bacterium. The whiplike strands at the bottom of the bacteriophage in Figure 3 are probably used to secure the virus to its hapless victim. But, it must be emphasized, this correlation of structure and function was relatively easy, because here we are still dealing with relatively large organized structures. We are not yet down to the level of molecules, which lie even beyond the reach of the electrons surging through an electron microscope. What is beyond? What surprises will await us when we can visualize molecules? What beauty? What knowledge? What insight into the normal and abnormal functioning of the cell?

At the present time we must resort to indirect methods for glimpses of molecular structure. The physicist bounces X rays off molecules and with piercing ingenuity interprets from the pattern of the rebounding X rays how the target, the molecule, is constructed. The hemoglobin molecule, we are told, looks like a spiral staircase which got lost. (Figure 5 shows a reconstruction of such a molecule.) Unfortunately, at the molecular level the relationship between structure and function is not yet clear. We do not have enough information to make correlations. It takes years of work to take the X-ray photographs, to make the measurements on the photographs, and to carry out the arcane calculations which yield the putative image of a protein molecule represented in Figure 5. It may take a decade or more to reconstruct the three-dimensional image of a relatively simple protein molecule such as insulin. So it is obvious that we are not even ready to start unraveling the structure of some of the molecular constellations which we know are the functional units in many biological processes.

At the present time, biophysicists are few in number and

they perform only on the periphery of the stage of biological research. However, many of us feel that relentlessly they will crowd out the biochemists from the center of the stage just as we have displaced the physiologists and histologists. When the biophysicists finally impose technological unemployment on us Zinsser's prophecy will be fulfilled. Biology will become applied physics. However, before that prophecy can be fulfilled there is still time—and need—for many more generations of classical biologists and even more generations of biochemists to wield their tools.

The effective participation by a new discipline in biology must await a definite sequence of development. Most of the fundamental observations in biology have been made by the classical biologists. The biochemists came later to explore the molecular mechanisms of the phenomena. It was in 1900 that the laws of heredity discovered by Mendel became generally known. But only in the early 1940s was the chemical pathway of the mechanism of heredity successfully explored: it was found that the genes exert their influence by the production of enzymes. The next problem which awaits solution, probably by the biophysicist, is, how does one speck of matter, the gene, induce the formation of another speck, the enzyme? When that will be answered is impossible to predict.

Could the development of biology be speeded up by inducing large numbers of physicists and physical chemists to become engaged in biological problems? That would hardly speed up the growth of biology any more than putting electricians on the job to start the electrical wiring of a building at the stage of the excavation of its foundation would speed up the completion of the final structure. A biological phenomenon must be recognized, explored, and developed before the physico-chemical aspects of it can be seen or approached.

Furthermore, not only must the field be prepared before the physical scientist can be fruitfully engaged in it, but the physical scientists themselves must be prepared. It takes years to train a physicist and then more years of retraining are needed in the handling of biological problems before he is useful to the field, and all of that costs money.

And that brings us to the vexing problem of the support of research. Until the Second World War not only was there inadequate financial support for the training of scientists in the United States, but there was precious little support of research altogether. Prior to the war most of the research in the biological sciences was carried on as a part-time (almost a hobby) activity of the professors at the universities and medical schools. Full-time research posts were scarce, and in most cases the incumbent in such a post had to have an independent income or had to take a vow of poverty. Furthermore, research programs had to be limited to very modest dimensions because of lack of funds for personnel and equipment. All too often promising leads could not be followed up. The ten-year lag between the discovery and the development of penicillin is perhaps the most vivid example of the tragic neglect of biological research. Fleming recognized in 1929 that penicillin might be an "efficient antiseptic," but he failed to concentrate this substance "from lack of sufficient chemical assistance." Had Fleming had a budget in 1929 of perhaps $8,000 a year for his project he could have succeeded with his program of chemical concentration. But in 1929 it was not easy, even for a distinguished and seasoned investigator, to obtain $8,000 for a research project in England or in the United States. Obviously, not every promising lead in research will yield rewards as rich as penicillin. But it would have been worth investing in a thousand fruitless searches to bring just one such as that to fruition.

It is not that our economy could not support more research. There was simply a lack of realization of the urgent necessity for such support. It is estimated that prior to 1940 our annual national bill for funeral flowers was $100 million. At the same time the total amount spent on medical research was $45 million, of which only $3 million came from federal government sources. Who was to blame for this lugubrious situation, which sounds as if it had been concocted by Jonathan Swift for one of his savage satires? The scientists themselves were partly to blame for the lack of both appreciation and support of their work. They had secluded themselves in their "ivory laboratories" and refused to communicate with the outside world. There was a time, not too long ago, when a popular article or book by a reputable scientist would cause considerable eyebrow raising. The reporting of the achievements of science was left to individuals from whose writings it appeared that the path to knowledge was illuminated by the flashes of genius of a few eccentric characters in dingy laboratories surrounded by a few pieces of odd glassware and a profusion of disbelieving, jealous colleagues.

Recently, the situation has been improved. There is extensive dissemination of knowledge by the scientists themselves. There is, also, better reporting of the work of scientists in the newspapers. To be sure, there are some research institutes and individual scientists who have shouldered a little too enthusiastically the burden of communicating their achievements to the general public, for, unfortunately, the knack for self-promotion and the talent for scientific research are not mutually exclusive abilities. This can do but little harm as long as we bear in mind that the headliners of today will not necessarily be the headliners in the history of science, and as long as we leave the evaluation of a scientist to his fellow scientists. Only an expert in the vast literature of science can tell whether an alleged discovery is new, only critical repetition can establish its truth, and time alone can measure its worth.

The support of biological research and of preparation for careers in research has been much improved since the Second World War. Three factors have contributed to our good fortune: the atom bomb, the Russian successes with their Sputniks, and the diseases of senators and congressmen. Unfortunately, while there is a lot of money spent on biological research by the federal government, the system of support is sporadic, undependable, and in many ways shortsighted.

It ought to be emphasized that most of us in biological research have nothing but admiration for the professional staffs of the government agencies subsidizing research. Since we all worship at the altar of private enterprise it is fashionable to denigrate the efficiency of public institutions. Actually the National Institutes of Health of the U.S. Public Health Service, the largest agency performing and supporting biological research is an extraordinarily competent organization. Its research staffs are second to none; indeed, universities keep raiding them constantly. Several heads of biochemistry departments in medical schools, including a Nobel Prize Winner, grew up professionally in the service of the National Institutes of Health.

The system of awarding grants to aid research in universities and private research institutes which has been evolved by the National Institutes of Health is also exemplary. Requests for aid are evaluated by panels of leading experts in the field, the majority of whom are not in government service. The panel arranges the projects submitted in order of decreasing promise. On the basis of such an order of merit as many annual grants are awarded as are possible from the funds made available by Congress. And that is where the trouble begins. For Congress determines not only the amount of money to be spent, which certainly is its prerogative, but more and more it is determining what line of research is to be followed. This trend is not new: purse strings are always too easily converted into bridles. The most promising paths to be followed are uncharted; but experts are brushed aside and congressmen become the guides. The philosophy of research of all too many in positions of power was epitomized by former Secretary of Defense Charles E. Wilson: "I don't want to know why potatoes turn brown." The remark was particularly deficient in both spirit and content. We know *why* potatoes turn brown. On the exposed surface of the potato enzyme systems oxidize the amino acid tyrosine, converting it into a brown pigment. For the complete knowledge of *how* the potato turns brown I would give my right arm. My willingness to sacrifice a limb for this knowledge is no mere cliché of exaggeration. With the knowledge gained in the trade I could grow myself a brand new arm without the spurs and creaky joints of the current one. For once we would know the ultimate mechanism of the functioning of just one enzyme, including how it is fashioned, and how it is controlled—we would be masters of the processes of growth, of regeneration, and of life itself.

Anyone unfamiliar with the history of the growth of knowledge or with the potentials and the limitations of science will tend to set an obvious, emotionally motivated goal for the scientist. Moreover, the paths chosen will be the well trodden ones which the layman in authority can understand. The recent history of the project for the search for anti-cancer drugs will serve as an example of congressmen and senators at work as directors of research.

The hope of finding a drug which will arrest cancerous growth has goaded many investigators into searches, however tedious, for such a chemotherapeutic agent. At its more inspired levels such searches have guiding principles or just hunches translated from basic studies of biological mechanisms to this hopeful end. At the lowest levels such searches are merely routine testings of chemical after chemical by drudging technicians. To increase the scope of the routine testing, the U. S. Public Health Service decided sometime ago to encourage private drug companies and some research institutes to engage in such projects at government expense. Buildings, equipment, and personnel for such ventures are all financed by the government through contracts made with the private agency. In 1957 $2,400,000 was recommended for this area in the President's Budget. Congress liked the idea and promptly doubled the amount. Next year the President's Budget had to back up the inflated program laid out by Congress the previous year. So $4,323,000 was earmarked for the program. Congress promptly trebled this to $12,371,000, and so the kiting of the contracted projects went on, until in 1960 Congress is allocating about $30 million to finance one year's hunt.

Thus, at the insistence of Congress a modest, sensible program of a routine search for a chemotherapeutic agent has become a vast project. This is a doubtful venture on many counts. There may be no better drugs than those already found, all of which are of only transient help against cancer. We find that cancer cells are so devilishly adaptable they can vitiate our efforts by developing enzymes to bypass the metabolic step made toxic by the drug. It is possible that the mastery of cancer will come only as a natural fruit nourished and ripened on a better understanding of the normal mechanisms of the cell.

Such knowledge is not likely to come from the vast army of technicians engaged in the routine testing of chemical after chemical. What we need is not more technicians engaged in pursuits a congressman can understand, but more first rate minds trained and reliably supported in their quest for knowledge. Yet the request of the National Science Foundation in 1960 for the modernization of the antiquated graduate teaching facilities in science in our universities was cut by the House of Representatives from $15 million to $2 million.

Someone has said that those who do not study history repeat it. It might be useful to recall the early history of two projects for which government support was sought. The story I am about to relate might be entitled "The Tale of Two Bombs." The practical feasibility of atomic detonations became apparent to several atomic physicists as soon as Hahn and Meitner published their discovery of the disintegration of uranium by neutron bombardment. The paralyzing fear of the physicists of the free world was that Hitler might beat us to the atomic punch. The story of their frustrations while trying to alert our government has often been told before. Briefly what happened is as follows: Dr. George Pegram, a distinguished physicist and then Dean of the Graduate School of Columbia University, wrote to the U. S. Navy Department introducing Dr. Enrico Fermi, Nobel laureate in physics and the world's outstanding atomic physicist. Fermi himself was to outline the possibilities of the construction of an atom bomb to the Navy. On March 17, 1939, Dr. Fermi got what euphemistically might be called an audition. The highest ranking officers the U. S. Navy could spare for interviewing a Nobel Prize winner were two lieutenant commanders! They heard him out, and then ushered him out.

After this rebuff five outstanding physicists, Enrico Fermi, Leo Szilard, Edward Teller, Victor Weisskopf, and Eugene Wigner decided in desperation to enlist the aid of Albert Einstein himself to reach President Roosevelt. Einstein wrote a letter in August, 1939, which was delivered by still another go-between in October, 1939. As a result of this letter, after prolonged study and conferences, the munificent sum of $6,000 was appropriated on February 20, 1940, to launch the atom bomb project.

Almost simultaneously a different bomb was clamoring for official attention and backing. Since this one was not proposed by a clutch of outstanding physicists including two Nobel Prize winners, it had no difficulty in getting the United States Senate to sponsor it.

Powdered carbon and liquid oxygen had been used as a blasting mixture in mining for a long time. Powdered carbon would be packed into a hole, liquid oxygen would be poured onto it, and a spark would detonate the mixture. During the early years of this century the digging of the Simplon Tunnel in Switzerland was aided by this explosive. For military purposes, however, a bomb with such ingredients was known to be useless. The manufacture and transportation of liquid oxygen to a mobile front presents insurmountable difficulties, and the mixture of powdered carbon and liquid oxygen is so frightfully unstable that the casualties among friends would be greater than among the foes. Moreover, it has no advantages over T.N.T.

It occurred to an inventor named Lester P. Barlow that carbon-liquid-oxygen mixtures used in aerial bombs might be our salvation in the Second World War, which was then raging. The ensuing fiasco sponsored by the Senate Military and Naval Affairs Committee is a long forgotten incident, but it should be salutary to recall it. All of the references I shall cite are from the New York *Times* of 1940.

On March 14 the Military and Naval Affairs Committee decided to meet to hear about the super aerial bomb. Four days later the Committee heard about the bomb that "destroys everything within miles." Mr. Barlow asked to be permitted to bomb livestock to compare his bomb with TNT. The Committee went through the ritual of burning its notes because the inventor had not yet patented. (Apparently no steps were taken by the Committee to plug up the Simplon Tunnel because it was dug with the aid of this same presumably unpatented explosive.) On March 19 the *Times* carried a news item expressing skepticism on the part of the War and Navy Departments. On March 20 it was announced that a demonstration was to be held for members of Congress and representatives of the Army and Navy. On May 17 the official test was postponed because the inventor left the site before the arrival of the Congressional members. Mr. Barlow claimed that the ordnance officers required him to have the bomb ready twenty minutes before the test and too much oxygen evaporated. (A bombing mission in a B-29 loaded with such bombs on a run lasting several hours would have been quite a picnic for the crew.) Finally, on May 26, came the hour of the test which was to shake the earth—or, at any rate, a small part thereof. Senators, congressmen, and reporters gathered at the ordnance proving ground in Maryland. Ninety-eight goats were tethered at distances from 200 to 1,000 feet from the center of the blast. A 1,000-pound bomb of 300 pounds of carbon and 700 pounds of liquid oxygen was placed on a pole thirty feet high. For the safety of the spectators the super bomb was detonated with a spark from a half mile away. When the distinguished witnesses reached what they expected to be a scene of utter devastation they found all the goats grazing contentedly. According to the New York *Times* the inventor said, "I am licked on it, but I had to try it to find out."

The story, however, was not yet finished. On June 18 Senator

Sheppard sought to obtain another $50,000 to back Mr. Barlow's experiments. It will be recalled that the atom bomb project up to then had received $6,000!

It is appalling that we have learned nothing in the past twenty years. In the area of science the advice of real scientists is still the last to be heeded. We are spending hundreds of millions of dollars on sending aloft a dazed man to circle the earth, even though sophisticated instruments can send back far more information than any man could, but at the same time the lower House of Congress in 1960 has cut by 10 percent even the Administration's modest request of $78 million for basic research for the National Science Foundation. This will mean that more than 60 percent of the meritorious proposals from three hundred different colleges and research institutes for some research projects will be rejected.

Who knows but that we may be starving some study which may yield the first clue into the origin of cancerous growth. Not only may not a congressman recognize such a study, the scientific profession as a whole, or indeed the scientist making the observation himself, may not recognize it. The title of the grant request would not be "Studies on the Cause and Cure of Cancer" but, more likely, it might read: "Studies of Anomalous Electron Resonance Barriers on Macromolecular Surfaces Derived from Heterocyclic Moieties."

Former Defense Secretary Charles E. Wilson's second aphorism on basic research, "Basic research is when you don't know what you are doing," is apparently shared by many in our government. Of course, the researcher knows what he is doing, but the Defense Secretary or the congressmen may not. The lack of appreciation by men of affairs who like to manage people has been plaguing scientists for generations. When Michael Faraday was asked about the usefulness of his discovery of electrical induction—a discovery which eventually created our electric and electronic civilization—he testily replied, "Of what use is a newborn baby?"

Men with managerial bent like to have a tidy picture in front of them: a simple goal set, the path to the goal outlined, and the outcome guaranteed. But tidiness is a minor virtue and simple goals are the yearnings of pedestrian minds. We would still be trying to convert base metals into gold and cure every ailment with poultices and blood-letting if during the past few centuries the creative imagination of all scientists had been shackled to goals set for them by their masters, whether bishops, kings, or congressmen. Since the federal government underwrites more than half of the scientific research in our country it is no exaggeration to state that our progress in science and its eventual technological applications depends upon the resolution of the impasse between the managerial mind of the congressman and the creative mind of the scientist.

The National Institutes of Health made a clever move a few years ago to close the gap between the Institutes and Congress: they renamed themselves! The Institute of Microbiology, for example, was renamed the Institute of Allergy and Infectious Diseases, and thus brought within the ken of a congressman. This is a deft concession which does harm only to those of us who must correspond with the Institute. But should we not as a nation be too grown up to have to resort to this sort of thing? Have we learned nothing from history? Practical applications, be they in medicine, technology, or warfare, only grow as a fruit from the roots of basic knowledge. Starve the root today, the fruit will wither tomorrow.

In biology and medicine the horizons where knowledge can lead us are still beyond our current vision. Who can say what is man's ultimate potential? That man's height has been in-

THE BEST IS YET TO BE

creased with our knowledge of nutrition is obvious, that his life expectancy is steadily lengthening is a firmly established fact. What is our unreached potential in the emotional and intellectual sphere? How much is it worth to us to find out? Could we afford to spend for this basic quest two percent of what we are spending in preparation for war? The reader—and his congressman—will have to decide that question.

We as a nation must decide on a long-term basis how much we can afford to spend on the search for basic knowledge. Only this way can a career of research be as dependably supported as say a career as a ladies' hairdresser. The benefits derived from the latter profession are obvious, therefore a ladies' hairdresser is well supported by our society. His income is adequate, he has job security; his equipment is elaborate and he works in a well appointed, air-conditioned emporium.

Too many American scientists work under conditions no self-respecting hairdresser would accept. Contracts on Government supported research are guaranteed for only one year; some young scientists are like migrant workers, going from project to project. Conditions are crowded and many laboratories have the attributes of slums.

Nevertheless many scientists resist the higher salaries and superior working conditions offered by industry or the much higher incomes in private practice for those with M.D.'s. For once they experience the sweet lure of the last frontier they cannot escape. The real researcher works because—to paraphrase some composer—he is less unhappy when working in his laboratory than at any other time.

The scientist is well aware, with Pascal, of his relative place in the universe:

For after all what is man in nature? A nothing in relation to infinity, all in relation to nothing, a central point between nothing and all, and infinitely from understanding either. The end of things and their beginnings are impregnably concealed from him in an impenetrable secret. He is equally incapable of seeing the nothingness out of which he was drawn and the infinite in which he is engulfed.

But the biological scientist is steeped in the satisfaction of studying a well defined area. He is studying "the central point between nothing and all," the mechanism of life and, therefore, of man, and he knows of no more rewarding a way to spend a life.

INDEX

Acetic acid, 80, 93, 95, 152
Acetylcholine, 222-23
Acidic group, 80, 120
Acidity of tissues, 130-31
Acids: acetic, 93, 95, 152; citric, 62; lysergic, 219-20; oleic, 246; phosphopyruvic, 61; phosphoric, 52; pimelic, 48; see also Amino acids; Fatty acids; Nucleic acids
Adaptation, mechanism of, 135
Adapted cells, 31
Adenine, 52
Adenosine triphosphate (ATP), 65-68, 92, 231
Adrenalin, 111, 221
Aging, 8-9, 34, 55-56; see also Longevity
Alanine, 107-8, 113, 123
Albinos, 36
Alcohol: ethyl, 151-52; fermentation of sugar into, 18-19, 38-40, 45; hydrocarbons as source of, 233; water formation from, 9
Aldehydes, 233
Alimentary canal, 53-54; butter in, 80-82; poisons in, 149; proteins in, 100, 157; sugar in, 60
Alkali, 81
Allantoin, 177
Allergies, 156-57
Aluminum, 2, 5n
Amines, 44; see also Ammonia
Amino acids: alpha, 237; beta, 237; carried by blood, 130; complexes excreted by bacteria, 165; as components of proteins, 10, 12, 16-17,

116, 203, 245, 246; isotopes and, 84, 91, 96; Krebs cycle and, 62; sequence of, 117; source of, 233-34; stability of, 236-37; structure of, 118-26; synthesis of, 182; water and, 59; see also specific acids, e.g., Glycine
Ammonia: in alanine, 108; amino acids and, 108, 113, 233, 234; and beginning of life, 59, 233-35, 238; in Pasteur's fermentation experiment, 39; in urea, 20-21, 111
Ammonium cyanate, 4
Anatomy, science of gross, 244-45
Anemia, 135-36; sickle-cell, 143-45; see also Pernicious anemia
Animal experimentation, 75-76; see also Experimental animals
Animals: body heat of, 60; hibernating, 112; light-producing, 67; marine origin of, 59, 141; need of water, 9-10; study of stomach of, 13-14; urea found in urine of, 3-6; use of ATP, 67; vitamin needs of, 51; see also Experimental animals
Anthrax, 155-56
Antibiosis, 161-68
Antibiotics, 161-68, 207, 246
Antibodies, 130, 143, 157-60
Antigen, 158-60
Anti-invasin, 34-35
Antimetabolites, 246
Antivivisection societies, 75-76
Arrhenius, S. A., 229-30